Sustaining Lake Superior

Sustaining Lake Superior

An Extraordinary Lake in a Changing World

Nancy Langston

Yale

UNIVERSITY PRESS

New Haven & London

Published with assistance from the foundation established in memory of
Calvin Chapin of the Class of 1788, Yale College.

Yale University Press books may be purchased in quantity for educational, business, or
promotional use. For information, please e-mail sales.press@yale.edu (U.S. office) or
sales@yaleup.co.uk (U.K. office).

Set in Monotype Bulmer by Newgen North America.
Printed in the United States of America.

Library of Congress Control Number: 2017933629
ISBN 978-0-300-21298-3 (hardcover : alk. paper)

A catalogue record for this book is available from the British Library.

This paper meets the requirements of ANSI/NISO Z39.48-1992 (Permanence of Paper).

10 9 8 7 6 5 4 3 2 1

Contents

NINE

Climate Change, Contaminants, and the Future of Lake Superior 216

Preface

WHILE I'M SITTING ON THE cliff over Lake Superior drinking my morning coffee, a pileated woodpecker comes for a visit. The bald eagles nesting nearby perch on an iceberg drifting down from the winter's ice pack. One eagle lifts off and dives for a lake trout, and a swarm of ring-billed gulls shrieks and mobs her, trying to drive her away from a gull colony that's expanding along the cliffs. Eight young loons paddle by, practicing their calls. Last spring, my neighbors were arguing over whether that was really a mountain lion they saw the other night (doubtful). Sometimes I'm lucky enough to hear the howls of the wolves that are now denning in the county forest across the road. Bears are so abundant that they have become a pest.

Forty years ago, few people figured any of these critters had a chance up here. Deforestation and failed farming had destroyed the habitat of many birds and mammals, and the erosion that followed clogged tributaries and estuaries with pollutants and sediments, devastating fisheries. Following World War II, industrial production had boomed across the globe. Mines and pulp mills along the shore of Lake Superior had dumped their toxic waste into local waters. Distant industries had released toxic chemicals that had moved from their sites of production and consumption into Lake Superior, making their way into fish and then human bodies. By the 1960s, the lake was at a tipping point, with the possibility of irreversible pollution.

Much to everyone's surprise, Lake Superior has witnessed significant recoveries in my generation. Forests and many of their inhabitants have returned after the devastation of the lumber era. The toxic waste sites left after the paper and mining booms have been partially cleaned up. Lake trout — once nearly extinct — spawn abundantly in the lake once more, one of conservation's great success stories.

None of these recoveries is complete, to be sure. The new forests are very different from the forests that were logged so quickly, and invasive species threaten aquatic ecosystems. Local women still wrestle with concerns over how much lake trout it is safe to eat when they're pregnant. New mining developments threaten to slice off entire mountainsides. Above all, global warming is now changing Lake Superior more rapidly than almost anywhere else on earth, remobilizing contaminants that communities thought had vanished.

Sustaining Lake Superior asks, What can we learn from the conservation recoveries of Lake Superior over the past century as we face new challenges of persistent pollutants that are mobilizing with climate change? Communities around Lake Superior have long struggled to address pollution concerns, and local, regional, and international efforts met with significant successes in the twentieth century. The nature of pollutants has changed since World War II, but, nevertheless, exploring the success—and failures—of pollution control in the past can help us devise resilient strategies for facing the challenges of pollution in a globalized, warming world.

The largest lake in the world (by surface area), Lake Superior contains 12 percent of the world's freshwater, a resource of enormous importance for a world where the supply of clean, drinkable water is increasingly vulnerable. This single lake is big enough to swallow all the other Great Lakes with a couple of additional Lake Eries tossed in. Vast enough to create its own storm systems, it births rogue waves taller than the buildings along its shores. Nearly every year another kayaker or two dies of hypothermia after a capsize, lulled into complacency by the lovely scenery. The cold in the winter stops you in your tracks and the wind can knock you down. The deepest trenches in the lake have never been explored. Gordon Lightfoot's "The Wreck of the Edmund Fitzgerald" wasn't just a lilting folk song—it was a warning.

Lake Superior presents an interesting challenge for people fighting pollution. Environmentalists often portray it as remote, wild, and pristine, suggesting those attributes make it worth special protection. But this leads to a dilemma: if it's so clean, why worry about a few toxics? On the other hand, if you stress its contamination to win support for environmental protections, then the lake loses its special status, so why bother focusing on it? Environmentalists finesse this dilemma by arguing that Lake Superior's particular geographic context—it is huge, northern, extremely cold, and distant from industrial developments—means that it is still the least spoiled of all the Great Lakes, retaining much of its wildness and splendor. Yet the very characteristics that have made Lake Superior less dirty from conventional pollutants, such as sewage and industrial waste, actually make it more vulnerable to the persistent toxic contaminants that have mobilized across the globe since World War II.

Fewer local sources of contaminants no longer mean better water quality when pollutants are increasingly mobile. Because Lake Superior is so huge and has only one outlet, it has a retention time of nearly two centuries. This means that a drop of water stays in the lake for, on average, 191 years—and contaminants can as well. Lake Superior is extremely cold, with an average annual temperature of 39°F. The cold water and the abundant winter ice cover lead to relatively low evaporation. So when toxics carried by atmospheric currents from Africa, Asia, and the lower Great Lakes find their way into Lake Superior, they tend to stick around. Lake Superior, like other cold northern lakes, has become a sink for the world's most distant and toxic contaminants. Toxics long banned in North America arrive windblown from distant places. Toxics from the past lie buried in sediments, stirred up into the water column by storms and bottom-feeding creatures. Pollutants in the lake blur the boundaries of space and time.

Pollution—and concerns about that pollution—have a complex history in the Great Lakes. As soon as industrial development burgeoned in the region during the nineteenth century, people began trying to comprehend and control industrial wastes. I argue in *Sustaining Lake Superior* that some of the earliest efforts to control pollution worked surprisingly well, for they rested on understandings of natural resiliency that made a great deal of sense at the time—and still have much to teach us. Two ideas were key. The first core idea was "dilution is the solution to pollution," or the belief that toxics do little harm if they are sufficiently diluted. The second key idea was assimilative capacity, which refers to the ability of ecosystems to absorb and break down pollutants. While some scholars have derided these concepts as mere apologies for industrial development, I argue that they actually encompassed understandings of natural resiliency. They assumed that human activities were part of larger natural processes, and so damage from those activities could be healed by natural processes as well.

In the early twentieth century, industrialization intensified, breaking apart the ecosystem relationships that had lent resiliency and assimilative capacity to freshwater ecosystems. Yet ideas about pollution were slower to change. After beaver were extirpated from watersheds, those streams and rivers lost much of their resiliency and ability to break down pollution, but few observers recognized it at the time. When forests were leveled, the ability of trees to pump pollutants out of the air and hide them away in the soil diminished as well. But because few observers had realized at the time that forests had helped to break down industrial pollutants, even fewer observers recognized that deforestation had reduced the capacity of watersheds to lessen the harm of pollutants.

After World War II, new persistent mobile synthetic contaminants, such as DDT, toxaphene, and PCBs, were produced and released in extraordinary quantities, and Lake Superior, like other northern ecosystems, became a sink for pollutants that had traveled thousands of miles. In the late 1950s, new understandings of mobility and global interconnections began to change the conversation about pollution and its spatial relations to centers of development. Unsettling research from nuclear testing showed that the north was no longer a pristine, remote place protected by its distance from industrialization. Northern communities most distant from Pacific nuclear testing showed some of the highest levels of contamination, suggesting two key ideas: first, that certain pollutants could rapidly mobilize into distant spaces, and second, that dilution offered little protection when certain contaminants could biomagnify in organisms at concentrations millions of time greater than their concentration in water.

When local citizens complained about pollution from growing industries such as logging, pulp and paper, and mining, they were not ignored. Rather, governments eager for economic development partnered with scientists who believed pollutants essentially stayed in place, thus remaining local concerns that could be managed with local agreements. State and provincial experts could partner cooperatively with industry, encouraging the adoption of technologies that would contain pollution enough to allow jobs and communities to thrive.

But as the understanding of pollutant transport radically changed in the 1950s, community, industrial, and government responses to those pollutants had to change as well. Case studies within this book explore the ways new spatial understandings of pollution challenged two key industries: pulp and paper and iron mining. Local concerns became global concerns, and global concerns became local concerns. Governance institutions struggled to adapt, and those challenges persist, particularly for the Indigenous peoples around the basin who eat contaminated fish.

Lake Superior may seem remote, but its waters are intimately connected to the rest of the world. Atmospheric currents bring chemicals from China, and pressures to mine iron ore in the basin are driven not by local or national markets but by a boom in China's steel industry. Yet while the processes that shape contamination have global roots, the effects are profoundly local. Mountaintop removal mining for China's iron ore demands would devastate local wetlands that have sustained the Anishinaabe for many generations. The toxaphene from Chinese, Russian, and African fields accumulates in the fish that swim under my cliff and makes its way onto my plate. What is global—financial markets, building booms, industrial farming practices in places with few environmental regulations—becomes local in the most intimate ways as it accumulates within our watersheds and within our bodies.

Acknowledgments

IT SOUNDS HOKEY TO THANK A LAKE, but so be it: I owe my
greatest debt to Lake Superior, which sustains the world with freshwater
and fresh hopes. My husband and I first spent a stormy winter week in a
cabin on Lake Superior in January 2002, and we fell in love with the lake's
winter intensity. Since then I've spent as much time as possible on Lake
Superior and its shores, first as a summer resident in a tiny cabin near the
Apostle Islands and eventually as a full-time citizen of the Keweenaw Peninsula. I am extraordinarily fortunate to live along the shores of the largest
lake in the world.

I am also fortunate to have found such a wonderful community along
those shores. In the Chequamegon Bay, Gail Amundson, Amy Billman,
Nancy and Phil Moye, Jan Perkins, and Peter Rothe have been stalwart kay-
aking, cycling, and hiking buddies, always ready to explore another corner
of the lake. In 2007, my neighbors along Roman's Point, just outside the
Apostle Islands National Lakeshore, came together to protect a small piece
of boreal forest on the lake's shores from future development. We failed,
but in the attempt we came to know each other well, creating a community
that has sustained us since. On the Keweenaw Peninsula, my husband and
I have been generously made welcome.

My fellow members of the Lake Superior Binational Forum were a
pleasure to work with, and their love for Lake Superior inspired and encouraged me. Bruce Lindgren, former U.S. co-chair of the forum, never let
us forget our vision statement: "As citizens of Lake Superior, we believe
that water is life, and the quality of water determines the quality of life."

Many members of the Bad River Band of Lake Superior Chippewa
and its natural resources staff have been generous with their insights and
perspectives. In particular, Matt Dannenburg, Dylan Jennings, Patty Loew,
and Chairman Mike Wiggins have shared their love for the lake and their
passionate commitment to protecting Lake Superior. Staff of the Great

Lakes Indian Fish and Wildlife Commission have been equally generous with their time and insights.

Like all historians, I depend on archivists. Staff at the Minnesota Historical Society, the Thunder Bay Historical Museum, and the Wisconsin Historical Society offered me every possible assistance.

Conversations with various scientists involved in Lake Superior research helped me understand some of the complexities of Great Lakes pollutants. I would like to thank Casey Huckins, Lucinda Johnson, Charles Kerfoot, James Ludwig, Carol MacLennan, Alex Mayer, David Mladenoff, and Mel Visser for their willingness to explain their research to me.

Many of my graduate students have been passionate about the Great Lakes, and their research has deeply informed my own. Special thanks go to my PhD students working on Lake Superior over the past decade, including John Baeten, Jim Feldman, Sarah Middlefehldt, and Michelle Steen-Adams.

Lynne Heasley and Daniel Macfarlane sparked a wonderful discussion on boundary waters at their Kingston, Ontario, "Border Flows" workshop, with lively conversations between Jerry Dennis, Colin Duncan, Noah Hall, Andrew Marcille, Emma Norman, and Ruth Sandwell, among others. Colleagues and students in numerous seminars gave me thoughtful criticism and feedback at many stages of this project. In particular, I would like to thank students and faculty who attended my talks at Northland College in Ashland, Wisconsin; Montana State University in Bozeman; Whitman College in Walla Walla, Washington; and the University of Kansas. Part of chapter 6 emerged from conversations in a workshop on Extractive Industries and the Arctic organized by Arn Keeling and John Sandlos with assistance from the Rachel Carson Center and Memorial University. John Thistle was a wonderful coauthor as we hashed out the connections between iron mining in the Lake Superior basin and the Labrador trough. Students and faculty in the Center for Culture, History, and Environment in the Gaylord Nelson Institute for Environmental Studies at the University of Wisconsin–Madison were inspiring colleagues for many years.

A sabbatical leave from the University of Wisconsin–Madison helped me begin research on this book, with additional funding from the American

Philosophical Society. The King Carl XVI Gustaf Professorship from the Swedish government allowed me the luxury of time and resources in northern Sweden to develop my analysis. Scholars and staff at Umeå University in Sweden were a joy to spend an academic year with, and conversations over our frequent *fikas* helped me broaden my perspectives. A National Science Foundation STS Research Grant #R56645, Toxic Mobilizations: Iron Mining Contamination, funded much of the research for the iron mining sections.

My editor at Yale University Press, Jean Thomson Black, has been a pleasure to work with. I thank the Press's anonymous readers for their thoughtful, detailed comments on the proposal and the manuscript. Their criticisms and suggestions improved this manuscript immensely. Any errors that remain are entirely my own.

Above all, I am grateful to my husband, Frank Goodman, for his love, encouragement, intellectual camaraderie, and support—and his willingness to move from his farm in southern Wisconsin to the Lake Superior watershed.

I thank *The Extractive Industries and Society* for permission to include in chapter 4 several paragraphs that first appeared in John Thistle and Nancy Langston, "Entangled Histories: Iron Ore Mining in Canada and the United States," 3, no. 2 (2016): 269–277. I thank the University of California Press for permission to reprint several sections in chapter 6 that first appeared in Nancy Langston, "Iron Mines, Toxicity, and Indigenous Communities in the Lake Superior Basin," in George Vritis and John McNeill, eds., *Mining North America: An Environmental History since 1522* (2017). I thank the University of Calgary Press for permission to reprint several paragraphs in chapter 7 that first appeared in Nancy Langston, "Resiliency and Collapse: Lake Trout, Sea Lamprey, and Fisheries Management in Lake Superior," in Lynne Heasley and Daniel Macfarlane, eds., *Border Flows: A Century of the Canadian-American Water Relationship* (2017). I thank *Environmental History* for permission to use several paragraphs that first appeared in Nancy Langston, "Paradise Lost: Climate Change, Boreal Forests, and Environmental History," *Environmental History* 14, no. 4 (2009): 641–650.

Sustaining Lake Superior

ONE

Ecological History of the Lake Superior Basin

ON COLD WINTER DAYS, I TAKE MY environmental history un-dergraduates on a snowshoe midterm, tromping through deep snows high in the forests of the Keweenaw Peninsula, a rocky spine sticking into Lake Superior. We start behind my house in the headwaters of Lily Creek, now a Superfund site filled with toxic tailings deposited by the long-abandoned Franklin copper mine. We weave through recovering white pine forests, then we make our way across a meadow that once supported potato fields to feed the hungry miners. After crossing through a young aspen grove that is rapidly colonizing an abandoned pasture for dairy cows, we find our-selves 600 feet above Lake Superior.

From there, we can look down upon the largest lake in the world: 31,700 square miles, or roughly the size of Maine (figure 1.1). If the clouds cooperate, we are high enough to see across the curvature of the earth 55 miles to Isle Royale, the largest island on the largest lake in the world. Beneath us moves enough water to cover North and South America a foot deep.[1] Lake Superior sits at the head of the Great Lakes basin containing more than 20 percent of the world's freshwater; what happens within this basin can affect the entire Great Lakes system (figure 1.2). My students un-derstand that Lake Superior's fate will shape their future in a world where freshwater is becoming increasingly vulnerable to threats of climate change, pollution, and overconsumption.

Early European explorers viewed Lake Superior with as much awe as my students do. The American geographer Henry Schoolcraft wrote in 1820: "He who, for the first time, lifts his eyes upon this expanse, is amazed and delighted at its magnitude. Vastness is the term by which it is, more

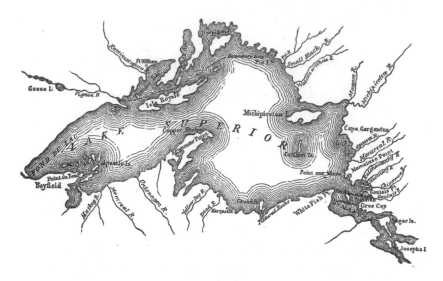

Figure 1.1 Map of Lake Superior, 1865. (Robert Barnwell Roosevelt, *Superior Fishing: or, the Striped Bass, Trout, and Black Bass of the Northern States* [New York: Carleton, 1865]. Digitized by the Library of Congress.)

than any other, described." In 1872 the Reverend George Grant added: "Those who have never seen Superior get an inadequate, even inaccurate idea, by hearing it spoken of as a 'lake,' and to those who have sailed over its vast extent the word sounds ludicrous. Though its waters are fresh and crystal, Superior is a sea. It breeds storms and rains and fogs, like the sea. It is cold in mid-summer as the Atlantic. It is wild, masterful, and dreaded as the Black Sea" (figure 1.3).[2]

Schoolcraft and Grant perceived Lake Superior as a near-eternal thing, unchanging over long time spans. But other nineteenth-century scientists, such as the Swiss geologist Louis Agassiz and the American ecologist Stephen Forbes, recognized that Lake Superior was a dynamic ecosystem changing over multiple time scales. Over centuries, as Forbes noted in 1887 in his classic paper "The Lake as a Microcosm," lakes have life histories just as people do. Even in the absence of industrialization, lakes typically age from a young, oligotrophic (low-nutrient) clear lake to an older, murkier, eutrophic (high-nutrient) lake. A lake's history often begins with the retreat

Figure 1.2 The Great Lakes watershed. (Drawn by Bill Nelson, modified from *Living with the Lakes: Understanding and Adapting to the Great Lakes Water Level Changes* [Detroit: U.S. Army Corps of Engineers Detroit District and Great Lakes Commission, 2000].)

of a glacier, which scrapes depressions in infertile landscapes that then fill with water. At first there's little organic matter to support much life, but shores and shallow areas capture colonizing organisms. When these die, they decompose and their dead bodies enrich sediments, so that soon more nutrient-demanding plants can survive. Plant and animal material decays and accumulates, supporting complex food chains. Growing fertility leads to more dead bodies, more sediments, more decaying things. Eventually a lake begins to age, shrinking as it fills with sediments. With less room for water, nutrients running in off the land are more concentrated. The lake becomes shallower, and when the weather cools in the fall, it is less likely to stratify into layers—which means those nutrients remain in circulation

Figure 1.3 Storms gather quickly on Lake Superior, leading to waves that make navigation treacherous.

throughout the lake all winter long, encouraging even more fertility.[3] Lake Superior is so vast that these aging processes happen on an extremely slow time scale, but cultural factors can accelerate any lake's aging.

Over billions of years, the Great Lakes have been even more dynamic. We are perched on one edge of the Keweenaw Rift, a rift valley as spectacular as the Great Rift Valley in Africa where humans evolved. Most of my students assume that glaciers carved out the Great Lakes. But their history lies much deeper in time. Nearly two billion years ago, where Lake Superior now sits was an inland depression filled with water. Iron suspended in the water fell to the bottom, forming bands of rich mineral deposits that supported a thriving mining industry in the nineteenth and twentieth centuries. But more than a billion years ago, the core of the continent began to pull apart, creating a 1,250-mile-long fracture curving across the center of North America: the Keweenaw Rift. Much of the rift has long been buried beneath sediments, but here on the Keweenaw Peninsula, thick layers of rock from the rift are still exposed.[4]

Nearly 1.109 billion years ago, a hot spot formed in the rift, much like what you can see at Yellowstone today, causing the crust first to rise up in a dome and then pull apart. Lava erupted from the spreading rift for perhaps 25 million years. Eventually the lava flows stopped, and the basin began to collapse under its own weight. Compressional forces developed, causing the fault blocks on the rift valley to reverse course and plow toward each other, thrusting over top of older sedimentary rocks. If this hadn't happened, leading the rift to fail, all of the North American craton might have sundered, creating a vast sea. But something stopped the splitting, and now Lake Superior occupies the basin of the deepest healed rift yet discovered. Along the Keweenaw Peninsula and on Isle Royale, the edges of the basin tilted up, exposing the Keweenaw fault, where rocks over a billions years old and rich veins of copper rocks lie close to the surface, easy picking for early miners.[5]

About two million years ago, the modern glacial period (popularly known as the Ice Age) began with continental ice sheets periodically expanding over much of the northern temperate zone. Each glacial cycle included about 100,000 years of ice growth, followed by 20,000 years of warming. Walls of ice more than a mile thick formed. Their vast weight pulled them forward, and they advanced, great sheets of ice crushing the land beneath them. As temperatures warmed during the interglacial periods, the ice retreated, leaving piles of debris behind, and great glacial lakes filled with meltwater held by dams of rock and ice. About 8,200 years ago, glacial Lake Agassiz burst free, and the release of the icy water into the ocean led to sea level jumps of over six feet in Europe's Rhine delta and probably clipped the Atlantic circulation, plunging the Mediterranean world into sudden cold.[6] What was profoundly local—glacial Lake Agassiz breaking open—had effects that crossed the globe.

In North America, the land rebounded when the ice retreated (figure 1.4). The northeast corner of Lake Superior, up in Canada by the little town of Wawa, continues to rise, rebounding almost eight inches in this

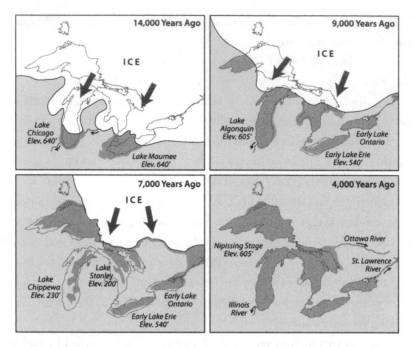

Figure 1.4 Retreat of the ice from the Great Lakes. (Drawn by Bill Nelson, modified from *Living with the Lakes: Understanding and Adapting to the Great Lakes Water Level Changes* [Detroit: U.S. Army Corps of Engineers Detroit District and Great Lakes Commission, 2000].)

century alone. This may seem very slow in human terms, but it's tremendously fast in geological terms. Nine thousand years ago, the Keweenaw Peninsula escaped the grip of the ice, but five thousand more years would pass before the northeastern corner of Lake Superior emerged. Today, changes much smaller create havoc with navigation, lakeside property, and infrastructure for shipping. We haven't adapted very well to the earth's longer term fluctuations, because within the time spans we can perceive, the lake seems remarkably stable.

Repeated glaciation scoured forests off the face of what's now the Great Lakes basin, so the forests that now exist across much of the northern tier of North America are young and still evolving as communities. Because people followed the melting ice north into the grasslands and new forests, human disturbances have been part of those forests for as long as they have existed—making it impossible to speak of pristine states for these communities. As temperatures rose at the end of the most recent ice age, forests began to expand. Rather than following the retreating ice, forests actually chased the ice up into the Arctic, for vegetation created a positive feedback cycle, increasing carbon dioxide, allowing grasses such as wild rice to became more fruitful, and attracting into the region Indigenous peoples who made the wild rice the center of their spiritual and economic relationships.

The forests that pushed back the edge of the retreating glaciers along Lake Superior ebbed and flowed with fires, insect outbreaks, wind storms, human pressures, and climate change. About six thousand years ago, as the climate continued to warm along the southern shore of Lake Superior, pines and hardwoods expanded into the region. Three thousand years ago, the climate cooled again and precipitation increased, leading to rippling changes in basin forests. Eastern hemlock invaded pine stands on rich, loamy soils in the southern portion of the watershed, and pines, aspen and birch persisted on sandier soils where hemlock wasn't able to thrive. On the north shore, where the granite of the Canadian Shield lay exposed by the retreating ice, a conifer forest known as the boreal forest developed on thin, acidic soils.

Boreal forests circle the northern hemisphere just beneath the Arctic, covering more than six million square miles—one of the world's largest terrestrial ecosystems. Ecologically, boreal forests grow in areas with very cold winters, short summers, and nutrient-poor soils. Conifers, such as balsam fir and white spruce, thrive in the overstory, while the forest floor is rich with berries. Beginning about 10,000 to 8,500 years ago, Indigenous peoples lived in the boreal forests of the Lake Superior watershed, shaping finer-scale forest disturbance processes and plant communities within the larger template of glacial history, climate, and soil. In the forests along the south shores of Lake Superior, the Anishinaabe cultivated gardens and wild rice lakes (figure 1.5). In 1840, Captain Thomas Cram noted that Indian women maintained gardens at the Anishinaabe village site near Lac Vieux Desert in the summer, with potatoes being particularly important.

Figure 1.5 Anishinaabe woman fanning wild rice, c. 1925. (Kenneth Melvin Wright, Photo 5151, Minnesota Historical Society.)

During the harsh winters, people dispersed in smaller family groups away from the summer village sites to hunt. When the days began to warm above freezing in the early spring, the Anishinaabe tapped sugar maples that grew in warmer sites for their sweet sap. Maple sap supplemented dwindling food stores, and maple sugar eventually became a key trade item with Europeans at fur trading posts.[7]

    The Anishinaabe developed aquatic resources within the basin as well as forest resources. Fish were widely dispersed and sometimes challenging to catch on the open lake, but their spawning cycles made them regular and predictable as a source of food. The Anishinaabe used Great Lakes fish not just for subsistence but also to sustain complex trade networks, first with other tribes, and then with European traders. They adapted a broad range of fishing technologies to the conditions they found in the Great Lakes, including spears, gaffs, hook and lines, and weirs in Lake Superior.[8]

In 1658, the Frenchmen Pierre-Esprit Radisson and Médard Chouart des Groseilliers ventured into the Lake Superior country, searching for their fortunes in beaver furs. Radisson wrote about their efforts to cross the Keweenaw Peninsula just two miles from where I sit today: "Some days later we come to the mouth of a small river where we kill some elk. We find meadows 25 miles square and smooth as a board. We go up the river 12 miles further and find ponds made by beavers. We must break the beaver dams to pass, and opening the sluice, it is wonderful to see the work of that animal which drowned more than 50 miles square and cut all the trees. Arriving at the river's source, we must drag our canoes over trembling ground for an hour. The ground trembles because the beavers have drowned great areas with dead water, whereon moss has grown two feet thick." To make their way over the portage, Radisson and Groseilliers followed the guidance of their Indian guides, who told them to "go like a frog," casting their bodies into the water and grabbing the moss to pull themselves along, drawing their canoes after them.[9]

Even given the propensity of writers to exaggerate their feats, this suggests a world shaped by beaver beyond our imagining. Across the continent, beaver lived almost everywhere there was water: from tundra to Mexican desert wetlands. In the most suitable habitat, they could build three hundred dams per square mile (usually, of course, there were far fewer). Perhaps 200 million beaver were present in North America before the fur trade began. More than 10 percent of the U.S. land area, according to ecologist Alice Outwater, may have been beaver-constructed wetlands when Radisson showed up, which translates to more than 300,000 square miles. When the beaver were rapidly pulled from the Lake Superior watershed and their furs sent to Europe, their old dams collapsed. Streams were no longer contained behind a series of ponds and dams. Water that had slowly seeped down to the aquifer now rushed to Lake Superior, carrying sediments, sands, and pollutants with it. Springs that had fed the watershed dwindled and water tables dropped, encouraging succession from ponds to meadows and then to grasslands. Outwater writes: "In a land full of beaver, the stillness of ponds and wetlands had allowed sediment to settle, clearing the water and providing a large reserve of nutrients that stabilized the ecosystem. . . . The beaver's wetlands had been home to a rich diversity of creatures of the air, land, and water, and without the beavers the fertility of vast areas was subtly reduced."[10]

Beaver were critical components of the hydrological cycle, shaping the assimilative capacity of watersheds through their ability to absorb and break down pollutants. Beaver defecated in the water, of course, so on a direct level they were sources of pollution, spreading giardia, known to trappers as "beaver fever." But indirectly, beaver were a key element in shaping stream resiliency to pollutants and floods. Assimilative capacity was to become an obsession for nineteenth-century pollution planners (discussed in the next chapter), who tried to maximize industrial development along the country's streams within the capacity of flowing waters to break that pollution down. But what planners hadn't realized was that when beaver were suddenly pulled out of the system, the assimilative capacity that had purified flowing waters lessened.

Cultural and ecological changes spurred by the fur trade transformed Lake Superior between the 1600s and the 1800s. Diseases and epidemics introduced by whites devastated native populations, and wars and famine exacerbated these processes. European invasions into the northern watersheds were a key tipping point in the history of Lake Superior, initiating feedback loops that "accelerated the transition of human and natural systems into novel states."[11] Wisconsin became a refuge for tribes fleeing the Iroquois expansion, and often several different tribes remained for protection in such locations as Chequamegon Bay on Lake Superior. As Richard White argues in *The Middle Ground,* the disruptions shaped white and Indigenous cultures alike, leading to new social relations and rippling effects on regional watersheds.[12]

The Agassiz Expedition

In the 1840s, as word of rich mineral deposits along the Keweenaw Rift spread, miners, military officers, and scientific explorers flooded into the region. As they ventured along the shores of Lake Superior, they recorded a world that they thought was whole, but one that was actually being fragmented by abrupt cultural and biological transformations. Louis Agassiz, the Swiss geologist who popularized the idea of glaciers and glaciation, was one of these explorers who changed the ways that Europeans viewed the upper Great Lakes country. Born in Switzerland in 1807, Agassiz moved to the United States thirty-nine years later. In 1848, he led a scientific expedition along the north shore of Lake Superior, paddling two canoes and one rowboat out of Sault Ste. Marie, at the east end of the lake, along Canada's north shore. His 1850 book about the voyage, cowritten with artist J. Elliot Cabot, is a rich source for seeing a lake in the midst of rapid change.[13]

When Agassiz paddled west into Lake Superior, he imagined himself an explorer entering new worlds, but he was dipping his paddle in waters long traversed by other men and women. The ancient peoples who came as the ice retreated, then the Anishinaabe, the fur traders and miners, and then the vacationers who followed them in search of a northern paradise — they all made the same complaints about hordes of black flies and clouds of mosquitoes, July freezes, drenching rains, snows that persisted until late in the summer. But Agassiz and Cabot were unique in their enthusiastic curiosity about a world in the making, a world in constant change.

Agassiz and Cabot were intent on finding evidence of glaciation and its changes. Near the Montreal River, they saw a polished and scratched rock that the professor "considered a proof positive of the correctness of the glacial theory. Its surface was a couple of hundred yards in extent, sloping regularly north to the water's edge." As they paddled west toward what is now Terrace Bay, they became even more excited, for they found evidence that the land had rebounded as the glaciers retreated. They wrote of a series of arced terraces "ascending one above the other to a height of several hundred feet . . . three main terraces . . . rising one above the other by steep slopes." Agassiz noted "the ancient terraces above the present level of the water," interpreting them as proof that the lake levels were once much higher. He wrote about the "mass of ice moving over the country" in a distant ice age, when temperatures were much lower than present: "such a mass of ice moving over the country would produce these effects of rounding and scratching the rocks," while the layered terraces marked "the successive paroxysms" of the ice as it advanced and retreated.[14]

Agassiz and Cabot recorded not just the marks of ancient history but also the evidence of recent cataclysmic social changes with rippling ecological effects. For example, they paddled past many abandoned mining locations, and they noted that close to most of them "the surrounding woods had been burnt, leaving the black stems, some standing and some lying crossed at various angles, like jack-straws. The ground was already covered with the fire-weed . . . striving to conceal the ruin with its showy blossoms. . . . In the course of the day we passed a deserted mining 'location,' marked by ruinous log-huts; and in another place we saw on the rocks the wreck of one of their bateaux." At another abandoned mining site they wrote that the marks of the miners' "labors, with the approaching winter before them, were everywhere visible. Wood had been cut and piled up; several log-cabins built and the cracks stuffed with moss and mud; and the paths through the woods showed where they went for fuel or to hunt. The ground was strewed with fur and bones of hares, and several lynx skulls were picked up by the men."[15]

They reached Michipicoten, a place now as remote as any along the lake. But in Agassiz's time, this was the center of a fur trading world, for as they wrote: "Michipicotin [sic] is the principal post of the Hudson's Bay Co. in this district. From it, the other posts are supplied, and the line of communication with Hudson's Bay passes through here. It is sixteen days' journey up Michipicotin and Moose Rivers to James' Bay." While they enjoyed the company at the Hudson's Bay post, they wrote that the wildlife had fled, including the caribou which were once found "all through this region" and now barely persisted within the basin, on the edge of becoming a ghost species that survived only in place names.[16]

Agassiz and Cabot described a conservation ethic among the Anishinaabe: "We were told that if an Indian finds a beaver-lodge, he cautiously traps a beaver or two, and then leaves them alone for the season, since otherwise the animals would forsake the place altogether. This he does year after year in perfect security that no one will meddle with them after he has proclaimed his discovery, and it is said that a beaver-lodge sometimes descends thus from father to son."[17] But already the fur trade "was very much on the decline," and fishing had begun to replace it: "Great quantities of fish are seined here; white-fish, lake-herring, trout, &c., not only enough for the use of this and other posts, but also some are sent down to the Sault for sale. The number of whitefish annually put up on the whole lake, Mr. Swanston estimated at three thousand barrels, worth on an average $5 a barrel. Of these, about one thousand barrels are sent away for sale."[18]

Indeed, the fur trade brought a commercial fishing industry to Lake Superior. The goal was not to feed the traders themselves but to replace corporate income that was declining as the beaver were depleted in the 1830s. Whites borrowed much of their fishing technology from local tribes, particularly the gill nets that were later blamed for devastating fish populations. Well into the twentieth century, Indian fishing technologies remained at the heart of the fishery; what destroyed the fisheries were not the Indian technologies (as contemporary observers claimed), but new national and global markets sparked by the fur trade.

Mining

Cabot and Agassiz traversed a world in the process of being commodified. The fur trade was fading and turning into a fisheries trade; the forests were on the verge of being logged to build Chicago; and the copper mining boom was accelerating. Along the Keweenaw Peninsula, Native Americans had long mined the rich copper deposits that the geologic processes of the Keweenaw Rift had brought close to the surface. Tribes traded these nuggets of copper and decorative art they made from the copper across the continent. Copper connected tribes who lived in the remote north to distant communities. Yet when whites came to exploit the same copper deposits, they brought with them very different conceptions of what copper meant: electricity, not art. As William Cronon argues for the Kennecott Copper mines in Alaska, because owners of new mining companies could conceive of land as capital, and because their own livelihoods did not depend on local resources, they could exploit land much more intensively than local tribes had done.[19]

On the Keweenaw, as in Alaska, a human population that depended almost entirely on local landscapes was invaded by a human population that depended on the local landscape very little. Before corporations owned by whites could exploit copper deposits in the Keweenaw, they first had to own rights to that copper, which meant displacing tribes. Treaties negotiated to pave the way for mining in 1830s and 1840s began the process of dispossessing the Anishinaabe from their lands.

Copper ore refining processes required huge amounts of water for the stamp mills—large machines that crushed the rock containing valuable copper. Water was returned to the lake contaminated with particles of copper-bearing tailings that filled bays, harbors, and inland lakes. By 1882, stamp mills were dumping about 500,000 tons of stamp sands into local waterways each year. The Keweenaw Peninsula near Hancock and Houghton was soon largely deforested to fuel the copper smelters, and it remained bare for three-quarters of a century.[20]

In the mid-1840s, the first of the iron ranges in the Great Lakes drainage basin came into production near Marquette, Michigan. Iron waste was less toxic than copper tailings, but the refining process added significant quantities of mercury to the watershed, soon becoming an important source of mercury to the lake. Some iron mines were vast open pits, others were deep-shaft mines, and both led to significant changes in fish habitats. Miners sliced off forests and the soils that sustained them to create the open-pit mines, leading to increased runoff and siltation in tributary streams. Deep-shaft mining pumped groundwater to keep the mines dry, lowering the water table and creating silt-filled runoff. Timber shored up shaft tunnels in deep mines, and the smelting furnaces demanded firewood. By 1903, for example, the iron furnaces of the Upper Peninsula used thirty acres of hardwood forest a day, every day of the year.[21] Mining-related runoff led to increased siltation that covered spawning beds, raised water temperatures, and changed river flows.

Logging

Loggers on the American shores of Lake Superior between 1890 and 1910 created new disturbances whose scale dwarfed the scale of earlier ones. Initially, logging was intimately tied to mining. Mines created the markets for

Figure 1.6 Logging a big load, northern Michigan. (Detroit Publishing Company. Library of Congress LC-DIG-det-4a03923.)

timber, and mines created the motivation for treaties that allowed lumber men to acquire title. Speculation in timberlands quickly followed. Between 1870 and 1909 the lumber industry boomed in Wisconsin, Michigan, and Minnesota. Sawmills in Wisconsin alone processed sixty billion board feet of lumber between 1873 and 1897 (figure 1.6). By 1898, the federal forester Filbert Roth estimated that only 13 percent of the white pine in Wisconsin was still standing. During the height of the boom, the states along Lake Superior provided "a third of the nation's forest products." But the boom was short lived. By 1939, the upper Midwestern states provided less than 3 percent of the nation's timber.[22]

The effects of this rapid deforestation on Lake Superior streams and estuaries were devastating. Harvest practices and slash fires increased erosion and siltation of streams, while splash dams scraped away riparian structure (figure 1.7). Tremendous quantities of sawdust were dumped in

Figure 1.7 Erosion following logging in the cutover counties of Wisconsin harmed watersheds and fish habitat. (Wisconsin Historical Society, WHS-3991.)

creeks and rivers, blocking fish passage. Stream temperatures rose as forest cover was removed, reducing spawning habitat for cold-water fish. Many contemporary observers noted how, as forests fell to the axe, lakes and streams became muddied, water dwindled in streams and wells, and floods became more common, along with summer drought. Roth wrote that deforestation had made "decided changes in drainage and soil moisture," diminishing the flow of larger rivers. Swamps had dried up, and hardwood thickets replaced wetland forests.[23]

Forests were the first great issue in the American environmental movement, and concern over logging drove the young movement. The devastation of logging in the Great Lakes states was a profound shock to many Americans. The white pines that had seemed as though they'd be around forever were gone in less than four decades. The lumber industry was migratory, moving quickly on to the next timber frontier, so it had little interest in future forests, regeneration, or sustained yield. Loggers left behind piles of waste timber ten to fifteen feet high that turned brown and dry in summer. Sparks set them alight, and intense fires followed. Immense tracts of clear-cut land went up in flames, burning so hot that soils were scorched, losing much of their ability to retain water. In 1908, the worst fire year in Wisconsin's history, 1,435 fires burned 1.2 million acres.[24]

As the timber industry depleted the Midwest and moved on, a new conservation discourse developed. If the nation continued to deplete its forests without thought of the future, it might one day find itself without the timber on which civilization depended. Foresters in particular were certain that, because of wasteful industrial logging practices, a timber famine was about to devastate America. In 1903, Wisconsin Governor Robert La Follette established a forestry commission, and the Wisconsin state forester Edward Griffith promoted a program of fire protection, reforestation, state forests, and tax reform. In 1911 the Wisconsin legislature approved a form of the Griffith program and appropriated funds to purchase cutover pine lands and establish a state forest reserve. But state acquisition of lands was controversial, and in 1915 the Wisconsin Supreme Court declared reforestation programs unconstitutional.[25]

European-Americans had expected that after the land was logged, it would be able to support their agrarian ideal: small farmers developing independent farms. It seemed only logical that if a piece of land could support tall trees, it would be fertile enough for intensive farming. Government and university experts in Madison, joined by the local press, merchants, and bankers, promoted small-scale farming on the deforested lands ("the cutover"). Between 1900 and 1920, lumber companies turned into speculators persuaded thousands of people from around the world to buy deforested land.[26]

Between 1900 and 1920, thousands of settlers moved into the region to farm the cutover. Although boosters promoted the north as a place of agricultural bounty, deforested land was not prime farm land. Because of the expense of extracting stumps and rocks, many farmers simply did not attempt to clear their land. Instead, they scraped by hunting, fishing, and working off the farm, together with limited dairying and growing of potatoes. The value of these small farms, usually only ninety acres and often not completely cleared, averaged only half that of Wisconsin farms in the rest of the state. Farmers did reasonably well during the 1910s and 1920s but struggled during the difficult 1930s and 1940s (figure 1.8). In 1914, during the height of cutover farm enthusiasm, state soil surveys classified only about 4 percent of Bayfield County as submarginal for farming. By the 1930s, the State Planning Board had reversed itself, classifying 75 percent of the county as submarginal and only 10 percent as good for farming. The land itself had not changed much in sixteen years, but perceptions certainly had.[27]

Attempts to farm cutover lands often ended in misery, and state and local governments in the cutover region struggled to adjust to decreased tax receipts. In 1927, the Wisconsin legislature authorized the establishment of county forests on lands that had been taken for nonpayment of taxes.

Figure 1.8 Farming the cutover, May 1937. Entire families participated in the back-breaking labor of building a farm on deforested land. (Lee Russell, Office of War Information, Farm Security Administration [Library of Congress LC-USF34-010960-E].)

Two years later, the legislature passed the Forest Crop Law, which reversed the 1915 Wisconsin Supreme Court decision and allowed the state to buy cutover lands for reforestation. In 1929, the state legislature authorized counties to adopt rural zoning ordinances, planning which lands could support farms and which should be allowed to return to forests. By 1940, all the counties in the cutover had adopted rural zoning, encouraging reforestation of the cutover counties. After World War II, agriculture dwindled in the area. Forests largely replaced farms across the region, in a massive state and federal effort at reforestation and forest protection.[28]

The cumulative ecological changes from the fur trade, mining, logging, and farming were profound. On the red clay soils of Wisconsin's south shore, erosion from farming was greater than erosion from logging. Nutrients bound to sediments moved off logged forests and the cultivated field that replaced them, clogging estuaries and streams. Clear bottoms became smothered with silt, which harmed spawning of cold-water fisheries. Bayfield Peninsula orchards, like other orchards throughout the Great Lakes basin, contributed toxic pesticides to the lake. Lead arsenates in particular were used at the end of the century, with significant effects on fish reproduction.[29]

While contemporary observers understood that these rapid changes might cause problems, it was rare to recognize that Lake Superior's geological context and history made the watershed particularly vulnerable to sudden ecological change. After the retreat of the ice, the Canadian Shield's thin soils and high resistance of its rocks to weathering had ensured that Lake Superior was biologically unproductive and slow to accumulate sediments. Lake Superior's geographic context meant that its waters were very cold, and that coldness shaped its ecology in profound ways. Like a few other cold, deep lakes, Lake Superior is ultra-oligotrophic, meaning that it is quite low in productivity and high in dissolved oxygen. In the summer, surface temperatures rise while temperatures below 600 feet remain at 39°F, and this variation in temperature stratifies the lake into three distinct layers: the epilimnion (the uppermost, warmest layer); the metalimnion or thermocline (the middle layer, which may change depth during the day); and the hypolimnion (the deepest, coldest layer). Twice each year the water column reaches a uniform temperature from top to bottom, and the waters mix.[30]

In most lakes, fish rarely use the hypolimnion because when organic matter decays oxygen gets depleted in the deepest layers of the lakes. However, in Lake Superior, low nutrient levels mean that populations of algae (and the animals that feed on them) remain low, so the water remains clear and dissolved oxygen levels remain high all the way down to the bottom. Lake Superior's coldness and low productivity mean that fish such as sis-cowet lake trout, which need substantial oxygen, can thrive in the hypolimnion, so deep that fishermen find it hard to reach them, giving the fish a measure of resiliency even when fishing pressures are quite high. But the particular ecological conditions that make Lake Superior excellent habitat for lake trout—cold, clear, and clean—also make it vulnerable to tipping over thresholds of sudden environmental change, such as a warming climate. If conditions warm, lake levels decrease, or nutrient levels increase, the hypolimnion may become depleted of oxygen, depriving cold-water fish of necessary habitat.[31]

Considering its enormous surface area, the lake's watershed is relatively small, which has historically helped minimize the contaminants that wash off the land into the water. But fewer sources of contaminants from the watershed have not always meant better water quality within the lake. First, the long retention time of Lake Superior means that a drop of water (and an associated contaminant) that enters the lake may remain there, on average, for nearly two centuries. Second, the cold temperatures of the lake and the structure of the lake bed mean that once contaminants enter Lake Superior, they may stick around near the shore for a long time, where fish can easily encounter them. In the spring, the nearshore waters of Lake Superior heat up more quickly than the deeper offshore waters. Because warm water is less dense than cold water, a thermal bar forms at the convergence of the nearshore water and the colder, denser, offshore water. The thermal bar acts as a barrier, concentrating floating debris, warm-water discharges, and pollutants within the nearshore area.[32]

Because of Lake Superior's geographic position on the Canadian Shield, lake depths sharply increase quite close to shorelines. This means that shallow, nearshore habitat is rare on the lake. Unlike Lake Erie, for example, where most of the lake is shallow, warm, and productive, only 4.7 percent of Lake Superior's area consists of nearshore habitat (here defined as areas where the water is less than 32.8 feet deep). In the nearshore, waves and current scour sediment from the substrate, maintaining good spawning and nursery habitat for many fish species while also providing excellent habitat for many aquatic invertebrates.[33] The relatively small area of nearshore habitat in Lake Superior means that fish that spawn in shallow waters—such as lean lake trout—are particularly vulnerable to toxics held close to shore by the thermal bar in spring.

Why does all this biophysical detail matter? In the late nineteenth and early twentieth century, paper industry towns such as Port Arthur (now Thunder Bay) were not oblivious to the potential problems of logging and pulp mill pollution in the lake. They knew that their drinking water usually came from the lake, and they also knew that the commercial fishing industry might collapse if pulp mill waste killed too many spawning fish. Early pollution discussions, however, tended to assume that the lake was one homogenous body of water. If you dumped a few gallons of toxics near the shoreline, surely that would quickly be diluted by the vast quantity of water in the lake.[34]

Yet Lake Superior's enormous size, which made planners hope that dilution might be the solution to pollution, actually worked against them. Lake Superior is large enough and cold enough that when thermal bars form, as mentioned above, they hold pollution where people and fish are more likely to encounter it. Fish also refuse to distribute themselves uniformly throughout the lake. They experience the lake as a complex set of interconnected ecosystems, not as a single large volume of water. During certain periods of spawning and fry development, they take refuge in the same places where pollution gets concentrated.

As the next two chapters on paper pollution argue, pulp mills and towns tried to manage pollution from growing industries, but their models did not account for the complexity of nearshore habitats, limnological conditions, bumpy shore bottoms, shoals that catch currents carrying sediments, or fish with minds of their own.

Industrializing the Forests, 1870s to 1930s

AREAS OF CONCERN (AOCS) ARE SITES along the Great Lakes that were designated in the United States–Canada Great Lakes Water Quality Agreement as suffering from the region's worst pollution. Strikingly, nearly all the AOCs on the Canadian side of Lake Superior come not from mining or the chemical industries, but rather from the pulp and paper industry. How did the pulp and paper industry—an industry that was intended to solve rather than create environmental problems in the Lake Superior basin—become the source of the region's greatest pollution problems?

As trees grew back on cutover lands, a new industry developed to exploit them. Pulp and paper forests supported a great deal of terrestrial wildlife, particularly species such as deer and ruffed grouse that thrived on young, patchy forests with lots of edge habitat. But while the young forests provided conservation value, aquatic pollution from the industry created a new set of pollution challenges that soon dwarfed the conservation problems presented by the lumber industry.

After the sources of valuable pine in Wisconsin and Minnesota were depleted, by the late 1890s, many American lumber companies headed west, but others turned north to exploit boreal forests along the Canadian shore of Lake Superior.[1] These boreal forests, dominated by spruce and fir, had not offered much merchantable lumber for construction or shipbuilding, so they had been spared intense clear-cutting during the late nineteenth century. But they did offer great opportunities for the young Great Lakes paper industry, which at the time was dominated by American corporations.

Both the American and Canadian governments saw the remote north as a hinterland in need of industrial development, and both nations saw trees as core to that industrial development. In both nations, mining facilitated logging by motivating treaties, creating markets for wood products, increasing pressure for railroads, and generating capital for industrial expansion. Yet with all the similarities between the two cases, distinct differences developed between the two nations' forest industries.

When the glaciers retreated from the south coast of Lake Superior, they left a sandy outwash plain with abundant clays from the old glacial lake. These soils grew rich stands of the tall white pines prized by lumbermen. But the clay soils were easily eroded after logging and farming, which meant that hardwoods such as aspen and birch sprang up after the cutover era. In contrast, along much of the Canadian shore, when the glaciers pulled back, they scraped the soils clean down to bare bedrock. That bedrock eventually supported a boreal forest dominated by spruce and fir. Neither spruce nor fir was much use for construction lumber, so the initial logging era bypassed the Canadian coast of Lake Superior (except for a few white pines growing along the lake shore). But when markets for paper developed, the Canadian spruce and fir forests became perfect sources for pulp.

The early paper industry initially had little use for the aspen and birch that grew back on the cutover lands on the American side of Lake Superior. Instead, the industry turned to the boreal forests, still largely unlogged, of northwest Ontario. The explosive growth of Chicago and the American Midwest meant that new paper markets were well within reach of northwestern Ontario, making it a "prime location in which to operate a pulp and paper mill compared with the north-eastern states and Quebec, the continent's traditional newsprint-making centres."[2] It was far easier for American mills to get Lake Superior trees to market via Great Lakes shipping routes than for mills in eastern Canada to transport that same wood overland.

American demand for pulp, American markets for newsprint, and American capital investments drove the expansion of the young Canadian pulp and paper industry. But before exploiting Ontario forests, American companies had to contend with Canadian desires for economic and regional development. Exports of unprocessed cut trees for American pulp mills would do little for Ontario job production.[3] If the Americans could be persuaded to invest in Ontario pulp and paper mills, then American capital could stimulate economic development in Ontario. A law forbidding the export of raw wood from Crown lands passed in Ontario, and in furious response, the United States placed a stiff tariff on newsprint imported from Canada. But American mills were unable to keep up with U.S. demand for newsprint, and by 1909 the price of paper began to increase sharply. Newspaper publishers pressed for a congressional investigation, and in 1909, Congress reduced the duty on newsprint, removing it entirely in 1913. The Canadian newsprint industry boomed in response, largely with American financing. In 1906 less than 1 percent of the newsprint used in the United States came from Canada but by 1916 nearly a third did.[4]

The growth of the paper and pulp industries in northwest Ontario was part of the larger growth of state power in Canada, similar to what Liza Piper has shown in *The Industrial Transformation of Subarctic Canada*. In Canada, wood for the pulp mills came primarily from public Crown lands, and Crown foresters were eager to stimulate the growth of industry to develop what they saw as a remote, primitive hinterland into a modern region. The original rules about forests had envisioned regulating only two uses for cut trees: firewood for fuel or timber for construction. Paper interests sought wood fibers, not straight, high-quality boards, so they cared little about mature forests or straight trees, let alone aspects of future forest quality. Thus they presented a new challenge for regulators.[5]

To comply with existing law, each company applied for what were called "timber limits," which meant control over vast expanses of public forests. Harvests on these public land were supposed to be regulated by professional foresters, who set the allowable annual cuts and other stipulations, such as required reforestation. Before investing in the infrastructure to cut and transport wood from Crown lands, the paper companies needed three key promises from the Crown: cheap prices for the wood harvested from Crown lands, sole access to those stands, and a promise of long-term rights to those stands. Canadian historian Mark Kuhlberg notes, "By the early 1900s, the newsprint industry recognized that each 100 tons of mill capacity required roughly 2,250,000 cords of pulpwood to sustain it on a perpetual basis."[6] Officials were prepared to meet these demands, negotiating cheap stumpage fees (the costs to cut pulpwood from Crown lands) and cheap water power rights. Canadian officials sent scores of letters to American pulp and paper mill executives, encouraging them to invest in forests along the Canadian coast of Lake Superior.

Decades ago, Canadian historians H. V. Nelles and Peter Oliver argued that Ontario's elected officials were little more than "client[s] of the business community." They did whatever they could to foster economic development in the region, providing natural resources to the industry at a fraction of their real value. Kuhlberg has complicated this argument by showing that Ontario politicians actually interfered with American pulp and paper barons such as the Minnesotan E. W. Backus, favoring Canadian lumbermen over the American-capitalized paper industry. Nevertheless, while Ontario politicians may have had mixed feelings about American investors, they promoted a united vision of industrial development in the hinterland. As Oliver argued, Ontario politicians had faith that large "units of production" would be in the best interest of "conservation, stability, and permanence."[7]

American companies reluctant to invest in Canadian plants were quick to exploit a key loophole in the Ontario pulpwood export laws. While trees logged off Crown lands could not be exported before processing, trees logged off mining and homestead claims could be shipped to American mills as raw logs. Investors found Ontario residents willing to register a significant number of mining and homestead claims along Lake Superior, and the American companies, nicknamed "timber pirates" by irate Canadians, exported the raw wood off those lands to American paper mills. As Bertrand writes, "From 1911 to 1920 . . . vast areas of our most accessible and finest spruce forests were staked out by these timber pirates." Eventually, on March 26, 1918, the Timber Act was amended so that no mining claims made after that date included harvest rights.[8]

American companies then targeted First Nation reserves because Ontario policy also allowed the export of pulpwood from Indigenous lands.

Wisconsin lumberman E. E. Johnson, for example, cut 25,000 cords off the First Nation reserve along the Pigeon River near Minnesota. American Charles Cox "secured the exclusive rights on all the spruce from the Indian reserve at Long Lac" then sold those rights to the Detroit Sulphite Company, making a profit on that single transaction of $60,000 —while the First Nations residents got nothing.[9]

As Americans continued to find ways to circumvent the intent of the law, Canadians feared that with American forests logged, Canada would become little more than a timber colony for American industry. Rather than simply saying "Canadian forests for Canadian jobs and Canadian profits," the Canadians tapped into a discourse of forest depletion and argued, We cannot have the forest devastation that happened in America happen here." In 1958, the Canadian paper industry executive J. P. Bertrand described a fellow timberman who "had a vast experience in the exploitation of forest timber from the northern states" and "was alarmed at the rate that this natural resource was being denuded. . . . Despairing of . . . a more orderly method of logging and a sound policy of conservation and sustained yield, he turned his eyes toward Canada." Bertrand added: "By the late 1890s, the vast forest lands of Wisconsin had been practically denuded of red and white pine. This great wealth was dissipated through shortsightedness and lack of a progressive forest policy."[10] In the early years of the pulp industry, paper industry executives positioned their industry as the conservation-minded solution to the environmental problems of intense lumbering, rather than as a new environmental problem.

Making Paper, Making Pollution

Although Ontario was convinced that the paper industry and forest conservation could coexist, water pollution from papermaking created enormous challenges for conservationists. A brief review of the technologies for papermaking will clarify why pollution became such a persistent problem. Wood is an abundant source of fiber, particularly compared to cotton and hemp, yet it took until 1843 to develop a profitable technological process for turning wood fibers into paper. The problem was lignins, chemicals that trees need to keep them tough, standing up straight, and resistant to rot. Turning trees into high-quality paper means removing as much of the lignin as possible without losing too many of the valuable long, thin cellulose fibers. Mechanical (also known as groundwood) pulping achieves this by ripping the cellulose fibers apart. It is an inexpensive process involving few polluting chemicals, but the paper is weak because the fibers have to be ground into short pieces, and lignin remains bound to the fibers. Newsprint, an example of a paper made with groundwood fibers, reveals the process's flaws. Abundant lignins mean that newsprint yellows with age as the lignins degrade, and the lignins also keep it from bleaching well.

Chemical pulping uses various powerful synthetic and natural chemicals to break down the lignins into smaller molecules that dissolve in water so they can be washed away from the cellulose fibers. The first chemical pulping process to be developed was the sulfite process, which uses strong acids to break down the lignins in spruce and other conifers with long fibers. Because the sulfite process can utilize conifers, but not hardwoods, it was ideal for paper production along the Canadian side of Lake Superior. The kraft (or sulfate) process, using soda (a base) rather than acids to separate lignins from fibers, was developed fifteen years later. The kraft process could take advantage of the short fibers of aspen, birch, and other hardwoods. The Canadian paper industry along Lake Superior was quicker to develop than the American industry because the earlier sulfite process could handle spruce and fir fibers.[11]

Pollution is integral to paper manufacturing, but the kinds of pollution produced vary with the technological processes involved. All trees contain phenols—natural plant chemicals that when unnaturally concentrated in pulp mill effluent can kill fish and disrupt hormonal systems. Pulp mills, regardless of type, discharge fibrous waste into waterways. Bacteria break the fibers down, using up oxygen in the process. (Sewage from municipal waste also gets broken down by oxygen-consuming bacteria.) Too many fibers dumped into streams too quickly means that the "biological oxygen demand" of the water skyrockets, killing fish that need oxygen to survive. In 1960, one scientist estimated that sulfite mills in North America discharged wastes equivalent to the biological oxygen demand of nearly two million people.[12]

The sulfite process for making paper discharges acidic effluent containing sulfur dioxide, a major cause of acid rain when it mobilizes into the atmosphere. Sulfur dioxide smells like rotten garbage, creating terrible odors near mills and causing irritation to eyes and lungs. Mercury is released as well, and benzene (a carcinogen), chloroform, methanol, nitrates, and ammonia. In contrast, the sulfate or kraft process doesn't release powerful acids, so it is often seen as less polluting. But it creates its own environmental challenges. Kraft mills produce atmospheric contaminants, some of which stink like rotten eggs — rarely a popular smell. Both sulfite and kraft mills release phenols into waterways, natural compounds from trees that are concentrated by the paper process enough to become toxic to fish downstream.[13]

Chlorine bleaching of pulp adds another key element of pollution: dioxins and chlorinated phenols. When lignins are removed from pulp by chlorine bleaching, the phenols combine with chlorine to make chlorinated phenols, for example, toxaphene and p-Chlorophenol. On their own, these are powerful endocrine disruptors, and they can also be contaminated with dioxins, some of the deadliest poisons in existence. If they're released into oxygen-deprived waters (streams just below pulp mills, where organic pollutants are being broken down by bacteria that use up available oxygen), anaerobic bacteria can methylate them so they become more likely to accumulate in fats and concentrate up food chains.[14]

The models for regulating pulp and paper pollution rested on nineteenth-century ideas of "assimilative capacity" — the belief that healthy streams had an innate ability to cleanse themselves of pollutants. This made sense for large flowing waters with small inputs of organic nutrients. Think of deer feces dropped onto the soil of a wild forest. When it rains, some get washed into the stream. Microbial communities in the soil and stream break the feces down, consuming oxygen in the process. As stream water flows over rocks and riffles, consumed oxygen is replaced by more from the atmosphere. Healthy streams have the capacity, in other words, to assimilate organic pollutants—hence the term "assimilative capacity." But if a feedlot or paper mill is placed along the river, the sheer volume of organic pollutants overwhelms the stream's ability to break them down. Bacteria consume so much oxygen that fish may become starved of oxygen and suffocate, resulting in massive fish kills, particularly during summer when river flows are low. State sanitarians tried to calculate how much organic waste a stream could handle over a given time period, so they could allocate the right to dump waste to competing industries. Assimilative capacity became a kind of property right, with industries asserting their property right to pollute a stream up to its assimilative capacity.[15]

The Great Lakes began absorbing pollutants from wood products as soon as the first sawmill went up on the shores of Lake Ontario around 1800. These sawmills dumped vast quantities of sawdust and wood scrap into nearshore estuaries and rivers, clogging harbors, covering spawning and feeding grounds for fish, and depleting oxygen as bacteria broke it down (figure 2.1). Samuel Wilmot, Canadian superintendent of fish culture, wrote an 1890 report that called sawdust "literally a poison on the spawning beds, a killer of marine insect and plant life on which fish depended for food." Wilmot added that sawdust "is an artificial product, alien to and engineering latent disease of various kinds, with fatal results in all waters where fish life exists."[16] Historian Margaret Beattie Bogue argues that the minister of marine and fisheries urged immediate action in the 1880s to avoid "serious permanent injury to the navigation and fisheries."[17]

The pulp and paper industry presented more complex challenges, for producing a single ton of pulp in the early twentieth century created over 80,000 gallons of wastewater containing lignins, cellulose fibers, and endocrine-disrupting chemicals. When bacteria broke down the organic materials, they consumed oxygen, making it unavailable for other aquatic life. A calculation called "biochemical oxygen demand" (BOD) emerged as the key measure of pulp pollution. Wood fiber lost during pulping, debarking, and milling all contributed to high BOD. Citizens living nearby were not happy and filed endless complaints about summer fish kills.

Industry and government alike, however, assumed dead fish and bad smells were temporary nuisances. As rivers flowed toward the sea, oxygen

VIEW OF LUMBER MILLS.

Figure 2.1 Sawmills in Washburn, Wisconsin, dumped sawdust and mill waste into the Chequamegon Bay of Lake Superior, filling in part of the harbor to expand the town on the old mill waste. (Wisconsin Historical Society, WHS-49029.)

(they hoped) would be restored and the rivers could assimilate new wastes. Before any new regulations could be placed on industrial waste, industry insisted that sanitary engineers needed to calculate the assimilative capacity of streams and classify North American waters by their most reasonable uses.[18] When American federal legislators organized a conference on stream pollution and introduced a congressional bill in 1934 to prevent the pollution of navigable waters, the chemical and paper industries blocked the bill, insisting "a stream's proper function was to provide for 'safe and proper disposal for measurable quantities of varieties of trade wastes.' "[19]

Assimilative capacity models made sense in the ecological conditions of the early nineteenth century. But as watershed ecologies changed with beaver removal, deforestation, dam building, mining, and development, watersheds lost much of their assimilative capacity. Even more important, new pollutants from pulp mills did not break down in water, so assimilative capacity models simply did not work for them. Many persistent toxics accumulate within organisms, particularly within fat. These chemicals can be magnified up food chains, concentrating as they move from microorganisms to little fish, little fish to big fish, big fish to carnivores like eagles and humans. The level of PCBs, for example, in the eggs of a fish-eating bird can be fifteen to twenty-five million times higher than the level in the lake water from which the fish came. Ideas about pollution, however, were slower to change than the pollutants themselves.

Regulating Waste

Much of the secondary literature on water pollution control is based within national boundaries, yet water and pollutants in the Lake Superior watershed flowed across those national boundaries. In both Canada and the United States, state or provincial governments had most of the responsibility for controlling pollution, with federal governments having less ability to regulate polluters. As geographer Arn Keeling writes about Canada, "Under the country's constitutional division of powers, provincial governments hold authority over most Crown land and natural resources, with some limited exceptions. This arrangement has historically limited the importance of the federal government in environmental and resource administration."[20] In the United States, until challenges after World War II, states similarly held most regulatory authority over water quality. But states, like provinces, faltered at controlling pollution, and when economic development and environmental regulation came into conflict, the environment usually suffered.

Geographic ideas about scale and distance structured pollution regulation. Before World War II, policymakers in urban centers typically believed Lake Superior pollutants were local concerns that could be easily contained in local spaces. Those policymakers perceived Lake Superior as a remote, distant place, and they essentially agreed with the modern Las Vegas advertising campaign: what happens in distant places stays in distant places. Additionally, they believed that by being situated so far away from most industrialization, Lake Superior communities were protected

from the worst sources of urban and industrial pollution. Finally, they assumed, as mentioned earlier, that "dilution was the solution to pollution."
The largest lake in the world, they believed, could handle the wastes from
the few industries situated within the basin. White elites who lived along
Lake Superior agreed with these assessments. They were boosters for their
region, and they typically felt a little pollution was a reasonable tradeoff for
economic development. They understood that pollution was happening,
but they believed federal governments had no need to get involved. Local actions, they thought, could contain pollution enough to allow water's
assimilative capacity to render it harmless while also allowing extractive
industries to thrive.

These were comforting thoughts, but not everyone shared them. For
the Anishinaabe, Lake Superior wasn't distant from anything; it was the
center of their home. Anything that damaged the integrity of the lake could
damage their health and culture. Fishermen also disagreed with urban arguments about pollution when they observed fish catches declining as industries expanded. Finally, new environmental concerns were expressed
by scientists, engineers, and unions after World War I. These groups may
have agreed with the implicit assumption that development was good for
the region, but they did not always agree with the assumption that pollution
could be easily contained.

Examining the expansion of Ontario's Great Lakes Paper Company
will illuminate some of these growing tensions. Great Lakes Paper, situated
in what is now Thunder Bay, Ontario, on the northwest shores of Lake
Superior, operated the largest pulp and paper operation in the world, em-
ploying some four thousand people. In 1919 Lewis Alstead and George
Seaman received an Ontario charter to incorporate. When they began
planning construction of the first pulp mill several years before receiving
the charter, companies needed a series of permits from local and provincial
authorities; they could not just start building an enormous pulp mill. The
town had to authorize the site, the use of water, and waste disposal into the
lake. The province had to authorize timber limits and power contracts for
energy development. These permits meant that the mill needed what soci-

ologists now call "social license" to operate, persuading key constituents that that the mill was in their best interest.[21]

During early permit negotiations, controversy regarding the project had already emerged. On March 10, 1915, one Port Arthur paper noted, "The new Minister of Lands Forests and Mines Hon. G. Howard Ferguson says that he is giving attention to the development of a pulp industry at the head of the lakes. I do not want this thing hurried. I will lose no opportunity to make the arrangements necessary to attract capital and provide a local market for both pulp wood and paper at that point, where I consider it is really needed." The notes from the town council meetings detailed that the $4 million to build the mill would come from the United States, but the company would be incorporated in Canada and supply jobs for 800 to 850 men. John Jay Carrick was mayor of Port Arthur beginning in 1908, and a member of Parliament from 1911 to 1917. The local paper viewed him with respect and a bit of wariness, calling him "a roaring lion seeking whom he may devour."[22]

In 1917, Carrick acquired the timber limits that became the mill's core pulpwood supply from the Crown, then he sold them to Alsted and Seaman in 1918. After making a tidy profit on the sale, he threw his political power behind the project, announcing in 1919 that construction would soon start in Port Arthur. But even with Carrick's support, the company struggled to get all the permits it needed.

The first requirement was a mill site, and rather than negotiate with the town council for a site, the company turned to the First Nations community at the nearby Mission Reserve. The public hearing record noted, "The Indians of the Mission Reserve at Fort William, in public meeting assembled, do hereby agree to surrender the lands described . . . at the price of $250.00 per acre, on the condition that one-half of the purchase price be paid to the Indians in cash."[23]

Water was the next order of business for the mill. The company first negotiated the "right to discharge waste liquid into Lake" and then "permission to pump water from Lake." The city resolved "that the City should also give every assistance it can to arrange that the Company may dispose of its waste liquor into Thunder Bay."[24] Although municipal permits for the site to take water and dump waste went smoothly, the company quickly ran into problems with hydropower contracts. How much power from the new hydropower plants along the rivers would be provided, and at what rates? Who would determine that? Power was central, and water was the source of power. While the company wanted cheap power, the growing town feared that monopolies on hydropower would drive up city rates, ultimately hurting its expansion.

Four years later, in 1923, the authorities were still arguing over the hydropower contracts for the mill. One advocate for the mill argued that by denying permits to the company, the town was wasting 4.5 million gallons of water every day. The agreement would give the city "a revenue from a portion of water which has hitherto gone down the river."[25] This is classic conservation discourse. Any water that ran into Lake Superior without first being used for hydropower was "wasted," its power lost forever. Great Lakes Paper investors wanted a "renewable ten-year contract which would provide a constant and continuous supply of power . . . to the mill site at the lowest possible price." Eventually, after a series of negotiations and scandals detailed in Mark Kuhlberg's history of Ontario newsprint, Great Lakes Paper finally got its power contracts, finished construction, and started operations.[26]

From 1927 to 1929 the company was acquired by notorious Minnesotan timber baron E. W. Backus, who built what became the world's largest newsprint mill next to the pulp mill. During the late 1920s, the paper industry boomed along Lake Superior, and the booster literature of Northwest Ontario celebrated its growth. One 1923 newspaper article was typical, stating that "the story of the astuteness of the financial men of the company in their successful fight to pull the company through that crisis would make a volume of romance in itself." What did companies need for those mills, besides the right kind of political deals? The main industry magazine of the time, *Pulp and Paper,* explained that the paper market depended on both water and forests, and while water was unlimited (it claimed), trees were not: "There seems to be plenty of water most everywhere and the supply does not seem to be diminishing. It does not have to be made. It just is or it is not. But paper has to be made . . . from a raw material which for all practical purposes is approaching exhaustion in one of the important markets of the world. The exhaustion is said to be due to a long permitted policy of exploitation instead of development."[27] The faith that water was essentially unlimited helps us understand how pollution expanded so quickly. If water is unlimited, then you don't need really to conserve or protect it: it will renew itself through assimilative capacity.

But forests were different; the industry feared that they could run out if a rational provincial policy did not protect the sources of prosperity. *Pulp and Paper* continued: "If a proper policy of reforestation is enforced there will be new crops coming on before the now-ready supply is half exhausted. It takes from 30 to 60 years for nature to produce a good pulpwood forest. Nature however refuses to take on the job unless given a chance. In fact nature needs assistance instead of reckless and careless opposition. Forests do not grow readily from ground covered by debris left by cutters who have only to do with harvesting an existing crop. . . . Neither can a crop be harvested after it is ripe if fire is permitted to ravish it."[28]

The demands on the forest to sustain a mill the size of Great Lakes Paper were tremendous. Between 1928 and 1933, Great Lakes Paper obtained 368,583 cords of wood, averaging $9.50/cord (for a total outlay of just over $3.5 million over five years). This translated into more than 28,000 acres cut each year. Company records state that "running the plant at maximum yearly capacity, there are cut over approximately 28,471 acres or 44.48 square miles to obtain 113,800 cords of pulpwood, which in turn produces 102,960 tons of newsprint."[29]

These are big numbers, but what did they mean for the forest's capacity? One way to approach this is to ask whether the harvests could be sustained at those levels. Local forests could not sustain these harvests, so the company negotiated timber limits farther and farther from their mill, adding to transportation costs significantly—and extending the ecological footprint significantly as well. By 1933, Great Lakes Paper controlled nearly 10,000 square miles of Crown forests. Harvesting 44.5 square miles a year to feed the mills was only about 0.4 percent of their holdings, and they had calculated they could harvest up to 2 percent of their holdings each year, in perpetuity. According to their calculations, they were well below their sustained yield capacity for the entire holdings.[30]

The 1933 company report on forest acquisitions continued: "As we have seen, our resources in standing timber are more than adequate for the perpetual supply of the two Great Lakes newsprint machines. . . . Therefore, with that in mind, it has been our constant endeavour to find new markets for other fibrous products. . . . No man or company can stand still and expect business to come to him or it. The position of our mill, possessing easy water connection with . . . Wisconsin and Minnesota, places us in a favourable position as a potential exporter."[31] From the perspective of Great Lakes Paper managers, they didn't need more restrictions on forest cutting; they needed more markets to exploit their timber limits.

Although company staff calculated that they could cut for centuries, by 1933, after just a decade of the paper boom, the forests and watershed of northwest Ontario's pulp landscape had been significantly altered. The timber limits closer to Thunder Bay had been largely cut over, as the forest acquisitions document admitted: "Almost all of these limits have been cut on, and much of the wood has been supplied to the Great Lakes mill."[32]

Harvests were sustainable only when expanding forest acquisitions were counted. Trees remained to be harvested, but the distances were becoming significant. Near the difficult-to-access Black Sturgeon limit, the 1933 document stated, "No wood has been cut as yet." However, in the view of the company, much of the forest's potential was wasted by nature herself, because 10 percent of the area had been recently burned over or was covered with wetlands and beaver ponds. Farther from the mill, in the Pic River limit (145 miles by boat), "thirty percent of the limit is waste, burn or water." Even as late as 1933, this area had not yet been surveyed or cruised for pulpwood, "but various estimates have been made ranging from 750,000 to 3,000,000 cords."[33] An estimate that varied fourfold was a rough one indeed. Nearly a third of this holding had recently burned or was a wetland, swamp, or pond. From the company's perspective, this suggested a forest of frequent disturbances that could eventually be rationalized with the tools of intensive forestry.

Diverting Water: The Long Lake and Ogoki Projects

The Great Depression brought hard times for Canadian pulp and paper mills. Overproduction led to a crash in newsprint prices and a severe contraction in the industry. Prices dropped for newsprint nearly 40 percent, plummeting from C$137 in 1921 to C$85 in 1930. In 1928, Great Lakes Paper temporarily suspended operations, laying off 300 employees, because of what the local paper called "the Newsprint Panic—closing mills and demoralized markets." Local papers assured each other that this was just a "temporary setback," but in 1931 Great Lakes Paper was placed in receivership. Backus's empire of forest, water, and power collapsed quickly, leaving behind a devastated community that struggled to find alternative employment during the Depression.[34]

In 1935, the Canadian scholar and politician Eugene Forsey wrote: "The pulp and paper industry has fallen on evil days since 1929. It is still our largest manufacturing industry both in value and output, and still ranks second only to wheat among the export staples on which our economic life is built." The global economic depression was partly responsible. But many of the problems were of its own making, according to Forsey: in particular, the overexpansion of plants in the 1920s and the overproduction of paper.[35] Yet rather than reduce mill capacity or production, in a Depression-era effort to rescue the pulp and paper industry, the Canadian government agreed to extensive re-engineering of Lake Superior water flows to create cheaper power for the mills.

Nearly every tributary running into the north shore of Lake Superior had already been dammed for power to supply the mills, devastating migratory runs of the coaster brook trout, now almost extinct, and other fish. Yet more forests and more hydropower potential lay over the watershed boundary. Although the problems the industry faced came from overproduction and depressed markets, not from too little wood, negotiations began about diverting great northern rivers across the watershed boundary into Lake Superior in order to float cut logs to the mill. The hope was that these projects would make the wood in those watersheds cheaper to transport and thus the paper more profitable.

During surveys of northern Ontario for railroads in the 1920s, planners had already begun discussing the possibility of diverting north-flowing rivers bound for James Bay into the Great Lakes. In 1933, the company forest acquisitions document urged rerouting the rivers in order to open up northern forests for paper. Because some of its desired forests lay in the James Bay watershed, and its mills lay in the Lake Superior watershed, hauling the wood over the watershed boundary was too expensive. Great Lakes Paper suggested "raising the water in Long Lake by a series of dams, and digging a diversion channel over the height of land between these two systems." The company did admit that "much more work would have to be undertaken before the feasibility and economic worth of the diversion channel could be finally settled."[36] The project would be far too costly for the company to undertake on its own, yet if provincial funds were poured into it, construction might go forward.

These discussions eventually led to an extraordinary project to serve the pulp industry in northern Ontario: the Long Lake and Ogoki diversions. Even now, these diversions continue to move 3.6 billion gallons of water per day—5,580 cubic feet per second—from a James Bay watershed into Lake Superior and from there into the entire Great Lakes system. During the Depression, the Ontario government "promoted the Long Lake Diversion as a method of providing employment, opening up a scarcely settled part of the province, and harvesting the virgin forest resources." Five Wisconsin paper companies—the Kimberly-Clark Corporation, the Hammermill Paper Company, the Mead Corporation, the Nekoosa-Edwards Company, and the Wisconsin Paper Company—formed the Pulpwood Supply Company, and the Ontario government contracted with them for the construction of the Long Lake diversion.[37]

The project went forward in in two phases. The first phase, completed in 1939 (just two short years after the agreement was signed) redirected the Kenogami River, a northward-flowing river, south through Long Lake into the Aguasabon River, which flows into Lake Superior. Workers dug a three-mile diversion canal through the height of land of the watershed and constructed two dams to control flows. Hopes of using the project to increase electricity production at Niagara Falls worried Americans, who feared for their own mills' ability to compete. The Americans initially blocked the hydro agreement, preventing Ontario from diverting water "in excess of that needed to transport pulpwood." But in 1941, "under wartime duress," the Americans backed down and allowed Ontario to use the diverted water for increased power production in the Niagara Falls area.[38] This in turn enabled the second phase of the project from 1945 to 1948, which included the Hays Lake Dam and Aguasabon Generating Station.

Even though the U.S. government had initially blocked part of the project to protect American corporations, those corporations ended up benefiting handsomely. Kimberly-Clark (based in Wisconsin) purchased controlling shares of the Pulpwood Supply Company and soon built a pulp mill and a town called Terrace Bay to house the workers and families. The project cost just over $1.331 million in 1939. Pulpwood Supply paid $300,000 of this; the province of Ontario paid $100,000, and Ontario Hydro paid 70 percent of the total cost, or $931,000. Between 1939 and 1974, Ontario Hydro power production net benefits amounted to almost $50 million (in 1974 dollars). Kimberly-Clark in particular did extremely well financially, saving between $63 and $68 million in log-handling costs alone.[39]

The pleasant company town of Terrace Bay boomed. Canadian federal and provincial governments encouraged industry partnerships to develop towns around a single industry: enormous pulp mills sited on the shores of the largest lake in the world. American firms infused funds into the region to develop the tremendous fiber resources of the boreal forests, particularly the long, thin fibers of black spruce. But not everyone benefited. First Nations communities were devastated both by the loss of fisheries and by their nearly complete exclusion from the project's benefits. In *Great Lakes Water Wars,* journalist Peter Annin describes the sad history of tribal loss of shoreline, sacred sites, graves, fisheries income, and land tenure rights. Changing river flows and inundation altered the Kenogami River, flooding out forested areas and desecrating an Indian Reserve cemetery, where graves were washed away. As planners in the 1970s noted, these problems reflect "present management strategies persistently favoring industrial uses." No consideration was given, the planners admitted, to "Indians' needs." Fisheries suffered when bark from log driving filled the river bed, consuming oxygen, emitting carbon dioxide and hydrogen sulfide, and harming spawning habitat.[40] Pollution concerns were almost entirely overlooked, even though commercial fishing groups were quick to express concern about the loss of fisheries.

Coping with Pollution: Cooperative Pragmatism

As concerns over pulp pollution's harm to fisheries grew in the early twentieth century, American and Canadian regulation relied on what historian Terence Kehoe terms "cooperative pragmatism." Cooperative pragmatism was based on "principles of voluntarism and informal cooperation, administrative expertise, and localism." Kehoe writes: "The engineers and board members in the Great Lakes Basin viewed water pollution as a relative concept and believed in setting waste treatment requirements on a case-by-case basis, taking into account a stream's waste-assimilative capacity, the receiving water's primary uses, economic considerations, and other local factors."[41] This approach allowed pollution, so long as other users' rights were also protected.

Sanitary engineers formed the intellectual core of cooperative pragmatism. With the rise of the industrial city and the enormous concentrations of sewage and garbage waste that resulted, the profession of sanitary engineering had become central to pollution control in both the United States and Canada, as urban historian Martin Melosi argues. Sanitary engineers didn't want to shut industrial development down to protect pure water. Rather, they wanted to find reasonable and efficient ways to keep drinking water from carrying infectious diseases. When downstream cities protested the contamination of their drinking water, upstream industries insisted that it would be cheaper to treat drinking water than to reduce waste dumping. Sanitary engineers agreed with industry, advocating municipal water purification rather than industrial effluent treatment.[42]

Industry groups, such as the Manufacturing Chemists' Association and the paper association, formed stream pollution committees, working with sanitary engineers from state or provincial agencies to develop

voluntary agreements that would clean up enough pollution to protect public health without burdening industry. The pulp and paper industry (like
the chemical industry that geographer Craig Colten analyzes) frequently
argued to state hygiene boards that they lacked the technological capacity
to reduce noxious wastes without bankrupting their firms. Industry typically delayed regulation with calls for more research. They would reduce
waste only after research led to technological processes that turned waste
into value.[43]

Delays were common in cooperative pragmatism, but the process
sometimes worked to protect public health while reducing pollution. Michigan offers an example. In 1922, Michigan sanitary engineers tried to curb
pollution from the Ontonagon and Munising paper mills on Lake Superior, but they met with little initial success because the companies insisted
they simply could not afford pollution reductions. But several years later,
research funded by the state and the industry led to ways of capturing and
reusing some of the lost fibers that created water pollution. At that point,
the companies were willing to work with regulators. In the 1927 issue of the
American Journal of Public Health, the sanitary engineer George Ferguson
reported that Michigan paper mills "state that they are already making arrangements to install machinery for the recovery of fiber as a logical business improvement. They emphasize the fact that they are doing this of their
own free will. However, when this subject was first brought to the attention
of the paper mill people in the fall of 1922, there was intense objection to
any change in their habits. After a little investigation they found that the
recovery of valuable constituents of the wastes interested them in a business way and they have been far more willing to cooperate in this problem
on that account."[44]

In the Michigan case, pressure from the state encouraged the industry to investigate ways to capture and reuse effluent fiber, thus increasing
profits while also decreasing waste and pollution into Lake Superior. Such
responses were rare, however, across the Great Lakes region. Archival
records from Wisconsin's pollution control boards in the 1920s and 1930s
reveal that regulators frequently met with little success, even when public opinion blamed the paper industry for fish kills. In 1925, the Flambeau

Paper Company of Park Falls, Wisconsin, was thriving along the Flambeau River (about fifteen miles south of the Lake Superior watershed boundary). The company dumped its mill waste into the Flambeau River; one release alone seemed to be responsible for killing between twenty-five and thirty tons of fish. Although earlier fish kills had been common downstream of paper mills, this particular effluent release sparked public hearings and citizen efforts to curb pollution. An analysis of this case shows how complex the efforts of cooperative pragmatists could become when new sources of paper pollution challenged assimilative capacity models. New regulatory authority over navigable waters in the state gave state agencies and citizens alike a tool to draw attention to pollution — but not a tool to regulate pollutants that watersheds could not assimilate.[45]

The Flambeau Paper Company owned dams along the river that it used to control water flow and power a paper mill. In 1925, the legislature had granted the state Railroad Commission "supervisory power over the navigable waters of the state, and control of the construction and maintenance of dams in navigable rivers," including the authority to grant permits to applicants to operate and maintain existing dams. With the new law, the Railroad Commission claimed the right to regulate pollution from pulp and paper mills if that pollution interfered with navigable waters. The company strenuously objected to this interpretation of the law, arguing that the legislature had never intended the state to exercise such power over corporations. Nonetheless, the Railroad Commission held extensive hearings in 1925 over alleged statute violations caused by stream pollution from paper mills.[46]

The Railroad Commission argued it had every right to "conduct an investigation and enter an order with respect to the discharge of obnoxious substances into navigable waters to the detriment of fish life."[47] Because navigable waters were necessary for economic development, the state asserted its interest in controlling pollution. But in the hearings, the state tried to claim a second, more tenuous, right to clean water: the right to recreational fishing. By insisting that the act "clearly establishes the right of the public to fish in all the navigable waters of the state," the state argued that the "right of navigation carries with it the right of fishing." Because

clean water is necessary for a healthy fishery, the state extended its naviga-
bility claims to arguments for water quality protections. Finally, the state
claimed a third right to regulate pollution: the right to protect the health of
its citizens. In testimony, the state sanitary engineer insisted that the "State
Board of Health is . . . given power to execute what is reasonable and necessary for the prevention and suppression of disease."[48]

To substantiate these claims, the state had a difficult set of tasks.
First, the sanitary engineers had to establish that they had jurisdiction over
state rivers and streams. Second, they had to prove that something in pulp
wastes injured fish. Finally, the State Board of Health had to establish that
dead fish and polluted waters might cause disease. Inherent in all this was
the tension of asserting regulatory authority over an industry that had become key to the state's economic development, particularly after lumbering
and farming had withered in the north. In other words, it was contested
authority, and the state had to tread carefully. The commissioner opened
the October 1, 1925 hearing by acknowledging that the commission's "jurisdiction in the premises is limited." Nevertheless, "by bringing out the
facts regarding stream pollution and its economic effect upon the state it
hopes to secure the cooperation of the paper and pulp mill companies in
a program that will ultimately do away with this pollution." This hearing,
the commissioner stated, "promises to be the most important one on matters pertaining to . . . conservation yet held by the Commission since it was
given supervision over the public waters of the state."[49]

The state Conservation Commission prepared a report in advance of
the hearings that attempted to quantify industrial waste entering streams.
The state found that "the greatest of all waste, is the liquor problem in
sulphite mills." The mills were wasting half the weight of the wood. Some
of the waste products were fibers that the companies were able to capture
and burn. Most of the waste, however, was a toxic liquid called "waste
liquor" that companies typically dumped into the state's flowing waters. In
one year alone, nearly four billion pounds of pulp mill waste had entered
Wisconsin's waters.[50]

An extraordinary amount of water was used in pulp and paper mills.
Most of that was eventually returned to local waterways polluted (figure 2.2).

Figure 2.2 Water pollution from paper mills in Wisconsin, 1955. (H. H. Bennett Studio, Wisconsin River, Wisconsin Historical Society, WHS-40826.)

The commissioner noted that "a mill making 130 tons of pulp and paper in 24 hours uses twenty million gallons of water in that time, which is almost as much as the amount consumed by an industrial city of 125,000 population."[51] The vast amount of water meant mill wastes were diluted, but pollution was reduced only if dilution rendered the pollutants harmless. The commission noted that new pollutants from chemical pulping processes made this comforting proposition unlikely. Because even the most dilute wastes were still harming public health and fish health, the problem of pulp waste was not *solved* by dilution but actually was magnified by it. This was a radical claim for 1925.

The problem that the inspectors faced was simple: nobody knew exactly what direct harm pulp wastes did to fish, much less what the indirect effects might be on food chains or reproduction. They did know that fish need oxygen, and "nature provided, clean water for fish with plenty of . . .

oxygen to breathe." But land use change had begun to unravel nature's assimilative capacity because the reoxygenation processes of natural streams were destroyed by the construction of dams and locks for pulp mills. Moreover, assimilative capacity relied on "scavenger birds and animals to patrol the lakes and streams and keep the water clean from pollution," and industrial changes had robbed Wisconsin of the wildlife that fostered nature's assimilative capacity. In modern terminology, the state recognized as early as 1925 that ecological changes were reducing resiliency.[52]

State sanitary engineer C. M. Baker acknowledged that no one should expect pure water; instead, the citizens of the state had a right to rational use. He testified at the hearing that "original purity can never be attained." Instead, the state needed to work with industry to ensure "the presence of free oxygen in water necessary to aquatic life." Biologist John Rue from the federal Forest Products Laboratory in Wisconsin testified on pulp waste's effects on fisheries, arguing, "In round numbers one and one-half million tons of sulfite pulp are made annually. For each ton of pulp approximately a ton of organic matter dissolved from the wood finds its way into the waste liquor, which, in almost all cases is dumped into the streams." In addition to waste liquors, spent liquors were produced after bleaching and then dumped into the rivers. They were particularly toxic, Rue testified, because the "solutions of chloride of lime" that were used killed fish eggs and the food that fish needed for survival.[53]

The companies didn't contest these arguments or deny that their waste killed fish (as they were later to claim). Instead, the industry simply stated that they lacked the technology to profitably reduce waste. If the state wanted paper jobs, the state had to accept pollution and dead fish.

Given the state's limited regulatory authority, the Railroad Commission and the Conservation Commission negotiated an agreement stipulating that their focus would be on cooperative research rather than on regulating discharge limits or enforcement. In the decision on February 20, 1926, the state found that "the discharge of industrial waste into certain streams is the only practical method of ultimate disposal in many cases." The Railroad Commission also reiterated the belief in assimilative capacity and self-purification: "A stream tends to purify itself by natural processes and will ultimately return practically to normal if the concentration of the wastes is not too great and sufficient time elapses before there is additional pollution." The state agreed that such discharge "constitutes a necessary and proper use of the stream" but only within limits, "provided that the dilution is so great as not to be materially objectionable as a menace to public health or interference with the natural aquatic life of the stream."[54] The trick was to determine those limits.

After the hearing, the state formed the Cooperative Committee on Stream Pollution, made up of industry members and state sanitary engineers. At its first meeting on March 31, 1926, the committee adopted a resolution acknowledging that "many pulp and paper mills in Wisconsin still have unnecessary fiber losses due to the fiber content of waste water discharged into streams." But the industry members of the committee insisted that sewage and municipal waste were worse than industrial waste, and "no practical and economical method of recovering or treating some of these wastes had yet been developed." So rather than enforce discharge limits on industry, the committee would "cooperatively seek a solution of the various problems presented."[55]

As part of the effort to understand the problem, in 1927 the state hired biologists to complete "the first modern scientific survey of Wisconsin rivers and streams," a 327-page report titled *Stream Pollution in Wisconsin*. The results were unequivocal: the rivers of Wisconsin were polluted, not just by municipal and domestic sewage, but also by industrial waste, and paper was the worst offender. Most important, the report argued that technology *was* available to reduce the polluting effects of mill wastes. A joint experiment in Park Falls showed that if a company temporarily detained the mill waste in a holding pond where it could be aerated, it was less likely to deplete oxygen downstream and kill fish. After the report was released, the Wisconsin legislature declined to require these aeration technologies to reduce mill waste. (Nearly half a century would pass before the industry adopted this technology.) Instead, the legislature created an additional joint research committee named the Committee on Water Pollution, giving it a mandate to research "economical and practicable" solutions to industrial discharges.[56]

In 1930, Wisconsin sanitary engineer L. F. Warrick reported to the industry in the *Paper Trade Journal* that their cooperative efforts meant that "progress is being made in the recovery and utilization of various chemical wastes which are the major source of stream pollution so far as pulp and paper mills are concerned." Warrick urged the industry to work harder to secure "satisfactory solutions for the many problems involved. . . . Though considerable progress has already been made under this cooperative program, the work has just started." In this public report, Warrick was polite, but in internal agency documents it is clear that he and his staff were already deeply frustrated. One of his staffers, A. Kanneberg, for example, noted that the American Paper and Pulp Association had failed to put up the funds they had promised for surveys of mill waste, concluding that after only five years he feared the "failure of the cooperative policy."[57]

In 1933, assistant sanitary engineer J. M. Holderby (Warrick's deputy) reported the results of the annual surveys of mill waste in the *Paper Trade Journal*. Holderby stressed that after seven years of research and cooperative studies, mill pollution was getting worse: "Waste flows, fiber losses and population equivalents per ton . . . were definitely higher in 1931 than in 1929." Some mills, Holderby argued, had reduced their pollution, which made it clear that the failure of other mills had nothing to do with lack of technology as they claimed. Holderby insisted on the need for "extensive improvement" in pollution control: "It is believed that the future of the industry in Wisconsin depends to a considerable extent upon the elimination of waste in manufacture and it will therefore, be of benefit to all concerned to reduce mill losses and the accompanying stream pollution to a minimum." Clean waters were increasingly important to Wisconsin citizens, and "public interest in the conservation and restoration of natural recreational facilities has grown hand in hand with an appreciation of their relation to the public health, welfare and comfort."[58] The industry's future in Wisconsin was threatened if the industry refused to control its pollution, Holderby implied. The public would not be patient with pollution forever.

Two years later, conditions were even worse because "fiber losses from the forty mills included have increased materially." Holderby insisted that technological processes were available that "will reduce the pollutional strength of the wastes at least 75 per cent," but mills were not yet required to adopt them. Fish kills had likewise not decreased. In a November 1934 meeting with industry, Holderby noted that, in spite of the industry's claim of an improvement in stream conditions, "as many or more dead fish complaints had been received by the Bureau of Sanitary Engineering during 1934 as had been received during 1933 and possibly 1932."[59]

By 1937, repeated fish kills downstream of the paper mills in the Fox River valley, where the Fox River entered into Lake Michigan, were angering locals. Commercial fishing businesses were closing, historian Frank Wozniak reports, and "massive die-offs of waterfowl at the Fox River mouth" were common. After commercial fishermen complained to the Wisconsin Conservation Commission, industry insisted that the problems stemmed from municipal pollution, not industrial waste. The state began a formal investigation in September 1938, finding that municipal and household discharges created only 20 percent of the oxygen depletion, whereas pulp mill discharges were responsible for the other 80 percent. In response, the paper industry announced yet another research effort that needed to be completed before regulation could be adopted: the Sulphite Pulp Manufacturers Research League.[60]

Fish kills continued. In 1941, Warrick warned the industry that commercial and sport fishermen were angry enough about pollution in the Green Bay area that they were trying to have a bill introduced into the legislature which would "prevent any pollution from sewage and industrial waste after specific dates."[61] If you don't fix this, he told the industry, the legislature may simply forbid any pollution release whatsoever, closing the mills entirely.

At the Committee on Water Pollution's meeting that same year, the vice president of the Flambeau Paper Company told Warrick that his company had finally developed a plan to reduce pollution: they would dump the waste sulfite liquor into a "worthless" swamp and let it settle for six months before discharging it into the river. The state engineers weren't impressed, pointing out that "a mill in California once tried such a project but that the liquor broke through the soil formation and reentered the river." A staff member from the State Board of Health warned about "the possibility of an offensive odor from the lagoon." The company requested additional time for research. The committee agreed, and the next year the problem was even worse, with the state reporting that "on the Flambeau River serious pollution conditions existed below Park Falls extending downstream as far as 35 miles."[62] Other sulfite mills had equally poor records.

The kraft mill owners insisted their effluents were much less deadly to fish than those released from sulfite mills. Yes, they contained toxic chemicals, but those chemicals were so diluted that they couldn't possibly hurt any fish. The Committee on Water Pollution sent a team of researchers to study this assertion. In September 1942 the researchers reported that they had to terminate the experiments because all the fish died when exposed to diluted kraft effluent. The kraft mill owners insisted that something else must have been the cause, but they refused to tell the state what was in the effluent. The scientists replied that whatever was in the kraft effluent, it was killing fish and possibly unraveling the entire aquatic food chain. They wrote: "The effects of pollution in the members of the food chain of the game fish should be investigated. This would include some protozoa, algae, crustaceans, mollusks, insect larvae, and minnows. . . . The effect of sulphate wastes in spawning and development of eggs and young fish should be studied. However, this is beyond the scope of the present experimental plant." Dr. C. A. Harper, a health scientist with the state, insisted that "it was a common sense approach to definitely ascertain the toxic ingredients and remove them from the wastes before discharge into streams." Even Warrick was frustrated and tried to persuade mills to tell the state what chemicals were in their effluent. The mills refused, claiming trade secrets.[63]

Winter fish kills were intensifying as well, raising new concerns about chemical toxicity from pulp waste. Summer fish kills made sense, for they fit within the assimilative capacity model. In the summer, water levels dropped (from power demands for mills), temperatures rose, and oxygen levels decreased, all of which decreased the assimilative capacity of streams. Winter fish kills, on the other hand, did not make much sense within the assimilative capacity model because the water was cold, water levels were high, and oxygen demands were relatively low. All of this suggested that fish kills came from new toxic chemicals that healthy streams could not break down—a frightening prospect for regulators and industry alike.[64]

The pulp and paper manufacturers pushed back hard when state regulators tried to require new pollution controls, arguing that they would simply leave the region if they were regulated. One report from the industry wrote: "Unremitting pressure from government for an ever-higher degree of pollution abatement makes top management increasingly reluctant to pour modernization money into an old sulphite mill. . . . All that has prevented permanent shutdown of some old sulphite mills is their owners' unwillingness to throw employees out of work and dislocate the local economy."[65]

Wisconsin state agencies expressed abundant concern about water pollution, yet they made remarkably little progress in reducing that pollution. Instead, frustrated regulators began to leave the government and go to work for industry, a revolving door familiar to historians of regulatory agencies. First Baker left the state to work for the pulp industry; then Rue left the Forest Service to do the same thing; and by 1941 Holderby left as well to join the industry's Institute of Paper Chemistry.[66]

In Canada and the United States, most reductions in paper and pulp mill waste before World War II came from new processes designed to increase economic output, not from effective regulation. Two technological developments helped save on materials and reduce pollution: new kraft pulping processes that lessened fiber waste, and new products such as chipboard that could turn waste fiber into profit, halving the suspended solids released into the lakes.[67]

States, provinces, and local communities in both the United States and Canada had responsibility for pollution regulation, and at these local levels, governments were reluctant to regulate industry when industry threatened to leave the region and relocate elsewhere. So, even though they stank and killed fish, most North American pulp and paper mills were barely regulated during the era of cooperative pragmatism. As Colten argues, even when the state or province tried to take action against the industry, "there was little effective pressure for manufacturers to abide by existing [pollution] laws before the late 1960s." Harrison notes that "Former B.C. Premier W. A. C. Bennett reportedly described the distinctive rotten eggs odour from pulp mills as 'the smell of prosperity.'" In Wisconsin, even with continued concern about fish kills and effluents, Wozniak writes that "water quality suitable for fish survival was not restored until the late 1970s."[68]

Citizens of both nations struggled with similar questions in the early twentieth century: When two uses of water conflicted, whose should have priority? When one user polluted the water in the course of gaining profit, who should pay to clean up that water? Should the polluter be responsible for ensuring clean water, or should the next user take that responsibility on for her or himself? And above all, at which scale of government should these questions be resolved?

Along the Canadian and American coasts of Lake Superior, governments encouraged industry partnerships to develop towns around a single industry: pulp and paper mills sited on the shores of the largest lake in the world. Lake Superior was remote from industrial centers, which presented a problem—distance to markets for forest products—but also an opportunity. The problem of remoteness could be solved by the lake itself, which offered cheap transportation to markets, at least when free of ice. Before interstate highways and cheap truck transport, industrialization required transportation networks, and the lakes were the way to get capital and labor into the heart of the continent and commodities back out. The isolation from markets presented an opportunity for regional economic development, with government and industry partnerships infusing funds into the region.

Lake Superior was critical for this vision: it provided the water for transportation to get the logs to the mill and the pulp and paper to markets. Its tributaries provided hydropower to run the mills cheaply. It provided the massive quantities of clean water needed for processing spruce fibers into pulp. And above all, Lake Superior provided the water for disposal of toxic effluents. For planners, Lake Superior seemed the perfect place to dump the wastes from the mills—dilution is the solution to pollution, experts reasoned. Surely, the largest lake in the world could handle the effluents from pulp production: mercury, dioxins, phenols from the natural plant chemicals, unnaturally concentrated.

For fifty years, paper towns along Lake Superior boomed: Marathon, Terrace Bay, Thunder Bay, Ontonagon, Munising. But, as the next chapter details, the human and environmental costs of intensive pulp production began to emerge soon after World War II. Anishinaabe communities were displaced from forests, suffering intense poverty and social displacement. First Nations communities in Grassy Narrows, Ontario, suffered mercury poisoning from the chlor-alkali plants needed for paper bleaching. Dioxin and PCBs created poison legacies that still confound the region. The paper and pulp industry brought three decades of economic growth that benefited many—but certainly not all—of the people living in the Lake Superior basin. Yet the pollution legacies from that boom era have persisted far longer than the economic benefits.

THREE

The Postwar Pollution Boom

AT THE END OF WORLD WAR II, new understandings of mobility began to change the conversation about pollution and its spatial relations to centers of development. It became clearer that the north was no longer a pristine, remote place protected by its distance from industrialization. New evidence about fallout from nuclear testing led to greater public understanding of the concepts of biomagnification and bioaccumulation, challenging the belief that dilution was the solution to pollution. Additionally, research in limnology helped shape new understandings of the complexity of lake habitats, leading researchers and communities to rethink the transport of key pollutants.

Yet these new understandings of pollution were accompanied by a postwar economic prosperity that rested on increased industrial production, which in turn meant increased pollution. During World War II, enormous volumes of industrial effluent were discharged into American waters. By 1947, industrial waste exceeded urban sewage.[1] Polluting industries boomed around Lake Superior in the years following the war, particularly the paper and mining industries.

North Americans took advantage of the economic boom by pouring into parks and wilderness areas, demanding new recreation opportunities. State, provincial, and federal governments were eager to comply with their desires for clean water, abundant fisheries, and easy access to beautiful nature.[2] How did governments around Lake Superior negotiate the increase in postwar consumer demands for clean water with a simultaneous boom in economic development? The cooperative pragmatists around Lake Superior hoped they had an answer: new chemical technologies that could produce cleaner lakes without threatening industrial production.

Aquatic Nuisance Program

Because the public wanted water that looked and smelled clean, and because regulators couldn't regulate industry, they decided to use copper sulfate, arsenic, and DDT to kill the by-products of industrial development, particularly algal blooms and the biting insects that loved them. Native fish that anglers didn't like could easily be killed with toxaphene and the lakes stocked with nonnative rainbow trout, a favorite fish for recreational anglers. These programs, ironically enough, sharply intensified exposures to persistent pollutants such as arsenic, DDT, and toxaphene—all in the name of clean water.

In 1904, two researchers with the United States Department of Agriculture published experiments showing that copper sulfate killed algae in irrigation reservoirs. Private individuals began pouring copper sulfate into Wisconsin lakes as early as 1918, hoping to reduce algal blooms that had "become notorious," fed by urban, industrial, and agricultural development that increased nutrient flows into lakes. When reports of fish kills from private applications of copper sulfate and sodium arsenate came to the state's attention in 1938, the Conservation Committee grew concerned. But rather than forbid the practice, the state tried to take control over its management. During the summer of 1939, the state Conservation Committee formed the Committee on Chemical Treatment of Lakes and Streams, "for the purpose of regulating chemical treatment of lakes and streams which had in the past been done by private individuals often with detrimental effects upon fish life."[3]

L. F. Warrick, the Wisconsin sanitary engineer who headed the state's public health program, was convinced that with careful scientific control, the state could use just enough copper sulfate and arsenic to kill algae without risking public health or fisheries. In a 1942 speech, Warrick told the American Water Works Association: "Public interest in clean waters has manifested itself in many ways in recent years. Pollution control agencies have been established in many states." He added that "the control of aquatic nuisances by chemical methods" will help with "the general problem of keeping waters clean and attractive." He concluded with a long discussion of why copper sulfate and arsenic compounds make good sense, even with their known risks as poisons. The sources of algal blooms were nutrients from sewers and trade wastes, Warrick admitted, but because we cannot control these nutrients, we must focus instead on controlling the nuisance.[4]

Warrick's own scientific staff working for the Conservation Committee were less enthusiastic. When his staff insisted that their experiments indicated arsenic hurt fish, Warrick urged them to determine the "tolerance level," where fish could survive the poisons but weeds could not. In 1945, the committee approved spraying twenty lakes with arsenic and copper sulfate and soon began to advertise its services across the state. A public brochure described the unpleasantness of the "aquatic nuisance" and the ease of chemical control: "Vile odors of large quantities of dead plant tissue may destroy tourist trade while resulting oxygen depletion will jeopardize fish populations. . . . Arsenical compounds are recognized poisons. . . . However, with proper control precautions no danger is experienced either to humans or animals." The following year, the committee warned the public of "common weeds that have produced nuisance conditions."[5] Many of these were native species that provided critical food for waterfowl, yet they were now classified as nuisances to be controlled with arsenic. The ideal water body was becoming little more than a waterpark, not a living ecosystem.

In July 1944 the state committee suggested the use of DDT to control midges and mosquitoes that bothered recreationists. The committee noted that DDT had "been restricted to the Army up until recent months." But the federal government had begun releasing small amounts for experimental purposes, so "effort is now being made by the Bureau of Sanitary Engineering to obtain a small supply of this new chemical to be used for experimental purposes, determining if the dosage of one pound per acre of marsh or pond area treated will in any way harm fish life." That summer, Wisconsin staff biologist E. L. Miller studied the effects of DDT on aquatic life, reporting back with disturbing information: concentrations of DDT "down to and even including 0.02 parts per million (ppm) will kill" fish, making it very dangerous to use. Miller warned the committee that DDT might lead to "rather severe upset in the entire biological community of a lake or stream treated with chemicals," with some changes appearing "over a rather long period" rather than immediately. Miller urged Warrick to avoid DDT and instead focus on the excessive nutrients that caused algae blooms "rather than to assume an entirely artificial method such as chemical control is necessary."[6]

Warrick overruled his staff scientists and decided to allow use of DDT in the state's rivers and lakes (figure 3.1). Initially, DDT was to be used only by state agents for research purposes. But this research program soon was the target of political pressures that the state found difficult to resist. The town of Oshkosh, for example, wanted to spray the "shore line of Lake Winnebago" with DDT from an airplane. After the state Conservation Committee denied the request, the local congressional representative, Congressman Keefe, called them in for a dressing down. The committee reversed its decision and allowed DDT spraying. That year's committee

Figure 3.1 Spraying DDT into Wisconsin waterways for insect control, date unknown. (Wisconsin Department of Natural Resources. Wisconsin Historical Society, WHS-60301.)

report noted, "While the concensus [*sic*] of opinion seemed to be that such wide-spread applications of DDT should not be attempted under present knowledge, it was decided to hold further action in abeyance pending further review of facts available on such treatment and study of the problem." This seemingly innocuous statement contained the history of failed pollution regulation in a nutshell. The state suspected that exposure posed serious risks—but it also knew those risks could not be proven with absolute certainty. What could be proven, however, were the risks of political interference and bullying. The burden of proof shifted from the user of DDT to show safety to the regulator to show harm.[7]

In 1947, the committee released a new set of regulations allowing DDT aerial spraying without a permit if the treatment rate was less than two pounds per acre. Fish-bearing waters did require a permit, and treatment at above two pounds per acre did as well. Even these limited restrictions were soon flaunted. A committee report from March 25, 1948, noted that one helicopter pilot was spraying the shoreline of fish-bearing lakes, streams, and cranberry bogs across northern Wisconsin with DDT, arsenic, and copper sulfate, all without a permit. By 1950, the committee discussed repeated instances of DDT air-spray operations without proper permits. At those same meetings, when staff scientists insisted on the need to study harm from DDT to fish, the department heads repeated their public assurance of "no adverse effects to fish."[8]

Even as these Wisconsin poison programs were booming in the name of clean lakes, the scientific field of limnology—the study of lakes—was flourishing at the University of Wisconsin–Madison. After the retirement of early limnologists Edward Asahel Birge and Chancey Juday, Arthur Hasler took the study of limnology in new directions. Rather than examining the "lake as microcosm," Hasler and his colleagues explored lakes within their larger environments. Hasler became fascinated by the cultural factors—for example, development, sewage, and fertilizers—that accelerated the natural aging of lakes. In 1947, he wrote in one of his key papers: "Enrichment of water, be it intentional or unintentional, is called eutrophication. . . . A survey of the limnological literature reveals a number of lakes in which entrance of domestic (cultural) drainage, over and above normal edaphic influence, was followed by marked biological changes." Hasler used a set of historical cases to argue that "fertilization hastens the extinction of a lake, erasing it as a human benefactor." He added that using chemicals such as copper sulfate and arsenic to treat algal blooms resulting from cultural eutrophication had only "brought additional problems" because they were "highly toxic" and "accumulative."[9]

Warrick agreed with the limnologists that development and industry were the ultimate sources of water quality problems. Yet he also knew his agency had no hopes of stopping development, thus chemical control appeared a sensible way of meeting the public desire for clean waters.

In the early 1950s, Wisconsin citizens began to question the safety of poison-spraying programs in public recreational lakes. At this point, Warrick's committee insisted to citizens what their own staff knew was untrue: that the poisons were completely safe for fish and people. When their authority was questioned by citizens, the agency staff dug in their heels, dismissing citizen concerns as emotional overreactions. By 1951, the minutes of every monthly meeting contained records of citizen complaints about the use of poison. The committee responded by insisting that spraying increased "the intrinsic value of a lake" while developing the "wise use of natural water" through "wise management practices." The 1953 minutes contain twelve formal complaints to the committee from various lake groups and fishermen's groups about the chemical spraying. By 1963, local lakeside groups opposed to spraying presented petitions to Warrick "containing 250 signatures of local citizens, objecting to the chemical treatment of the lake."[10]

As complaints about poisons sharply increased, so did the use of those poisons. The 1950s saw a near-exponential growth in arsenic spraying (figure 3.2). In 1949, in the Madison lakes alone, 3,178 pounds of sodium arsenate were applied. Additionally, between 1925 and 1950, on one small Madison lake, 1,697,639 pounds of copper sulfate were applied.[11]

Warrick's assurances of safety rested on a dream of precise control, the hope that technicians could calculate and then apply the exact amount of poison that would kill algae but not other life. But the programs themselves resisted control. Spray drifted into creeks. Hotshot pilots refused to

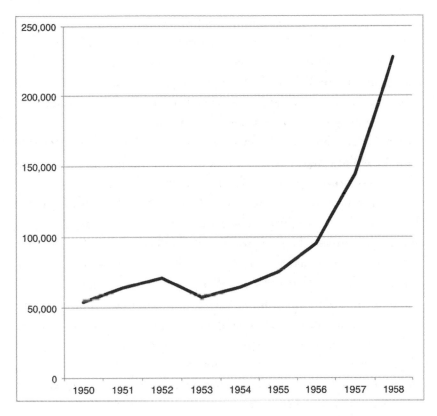

Figure 3.2 Pounds of arsenic applied to lakes for the "aquatic nuisances" program in Wisconsin between 1950 and 1958. (Data from Aquatic Nuisances files, Wisconsin Historical Society.)

bother with pesky permits. Homeowners figured if the state could spray, so could they, and if a little was good, more was better. But with each mishap, the committee was quick to blame the growing problems on environmental conditions—sudden rainfall, a bit of wind—rather than on their decision to use lethal chemicals in public waters.

The next step the state took was to kill trash fish (native fish that anglers didn't value) so that anglers could have more of their favorite sport fish, typically the alien rainbow trout. Anders Halverson, in *An Entirely Synthetic Fish*, has given us an excellent analysis of the enthusiasm for stocking with rainbow trout, and I won't repeat his detailed arguments here. But one example from a Wisconsin team in 1973 is worth offering. Robert A. Hughes and Fred Lee described the effort "to overcome fish population deficiencies" by eradicating native fish with "a chemical toxicant, such as toxaphene, and restocking with wanted species."[12] Natural conditions, in other words, were increasingly viewed as deficiencies to be corrected.

Beginning in the 1950s, toxaphene became the dominant fish-killing chemical. "Toxaphene" refers to a group of turpentine-smelling chemicals made from pine oil and chlorine. These two natural chemicals, byproducts of the forest industry, were combined into a synthetic substance that killed trash fish at lower concentrations than desirable fish. Toxaphene had first been described in 1906 by two chemists, G. B. Frankforter and Francis C. Frary, who were searching for profitable byproducts of waste from logging. Little attention was paid to the chemical until in 1947, when two state agricultural researchers from Delaware, W. Leroy Parker and John H. Beacher, reported on its powerful ability to kill insects. Agriculture research on the compound boomed. In 1947 alone, state entomologists and extension stations used over two million pounds of 20 percent toxaphene dust to test its effectiveness. Hercules Powder Company, incorporated in Wilmington, Delaware, built a factory in Mississippi devoted to making toxaphene. Hercules promoted toxaphene intensively. For example, one 1948 pamphlet the company aimed at farmers was titled "More Cotton, More Profit with Toxaphene" (figure 3.3). Hercules bragged that the "swing toward modern chemically-made insecticides" was the "biggest news for cotton growers today." "Cotton is a money crop!" sang the brochure. After Hercules's 1943 patent expired, 147 applicants got permission to sell 647 different formulations of toxaphene (making it difficult if not impossible to track its expansion around the globe).[13]

In the late 1940s, after "dead fish appeared in waterways adjacent to fields which had been sprayed or dusted with toxaphene," some observant fisheries biologists recognized that it might be a useful chemical to kill unwanted fish. Laboratory tests in the early 1950s showed that goldfish tolerances were below 0.005 parts per million—extremely low. Toxaphene killed trout fingerlings at equally low concentrations, but carp took ten times the concentration to die. Field studies showed that reservoirs

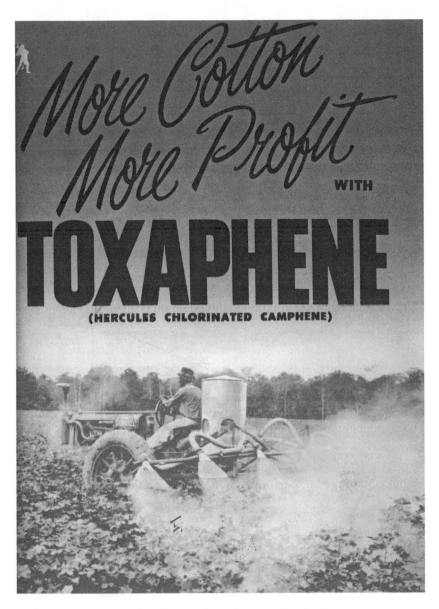

Figure 3.3 Toxaphene advertisement from Hercules Powder Company, 1948.

treated with toxaphene at 0.1 parts per million concentrations remained toxic for ten months. In 1954, the fisheries researcher Jack Hemphill found that populations of insects and zooplankton in treated lakes crashed, and a host of subsequent studies showed toxaphene's toxicity persisted long after treatment. For example, Michigan lakes were treated in 1949 and 1950 and remained toxic for thirty-three months. A 1962 study showed that fish swimming in the lake when it was sprayed accumulated toxaphene at up to 15,000 times the water concentration. Even more unsettling, fish planted more than a year after toxaphene had been sprayed accumulated toxaphene as well. Yet these early results showing that toxaphene was persistent and bioaccumulative seemed to raise little alarm in the fisheries community. Its use as a piscicide boomed, until it became the most commonly used fish-control poison in the Lake Superior basin.[14]

Why did the aquatic nuisance program grow so quickly in the name of conservation? In part, the "better living through chemistry" enthusiasm shared by many North Americans explains the willingness to spray the way to clean water. Nifty postwar chemicals were going to solve all our problems. Plastics such as bisphenol A would make happier, cleaner, shinier households; hormones such as DES would make bigger, better babies; pesticides such as DDT would create a cornucopia of shiny, perfect food.[15]

Even at the time, however, many scientists, citizens, and regulators resisted these promises. State regulators expressed consistent frustration with industry. Scientific research to show harm was abundant, and regulators frequently discussed that research in their committee meetings. Agency scientists prepared report after report showing that arsenic, copper, DDT, and toxaphene harmed fish at much lower doses than needed to control weeds. Yet those results and discussion rarely led to program changes.

Lack of regulatory authority over industry constrained regulators' ability to bring about the clean water they promised. They had control over only their aquatic nuisance program, not over the industries creating the conditions that made spraying seem necessary. They used the authority they had, trying to modify within increasingly narrow parameters the spray programs. Yet the more they discussed tiny adjustments to those spraying regimes, the less they challenged the basic premise: Was poison spraying necessary for clean waters? State agencies were not in a position to question economic development itself. The longer they served on committees with industry partners, the harder it became to question their shared assumption that pollution was a technical problem that could be solved by technical adjustments.

Changing Understandings of Dosage and Dilution

Arsenic as the answer to the demand for clean water and good fishing may seem absurd to us now. But it made sense within early twentieth-century models of pollution that believed bacteria from sewage were the worst pollutants. Geographer Craig Colten argues that sanitary engineers sometimes welcomed industrial polluters because toxic industrial chemicals could kill disease-causing bacteria. Well into the twentieth century "the perception that toxic wastes were beneficial additives to waterways with large populations of pathogenic bacteria" persisted.[16]

Understandings of pollution changed in the 1950s as new data from wartime nuclear and chemical testing became public. For generations, a core assumption had been that "dilution is the solution to pollution." With enough water to dilute discharges, toxics could be made relatively harmless. Several assumptions were inherent in the "dilution is the solution" argument. First, regulators had assumed that toxins would spread themselves evenly throughout a big lake, so that they would be diluted by the entire volume of water. Growing understandings of limnological complexity began to challenge this assumption in the early twentieth century as scientists were able to measure temperature differences and underwater currents that concentrated and transported toxics in unexpected ways.

A second key assumption was spatial. People had long assumed that if you were closer to a toxic discharge, it was more likely to hurt you. This makes intuitive sense. Explosions hurt more when you're close. Fires are hotter when you stand next to them. Raw sewage gets more diluted the farther you are from it. But these intuitive associations don't hold for many toxics, which move in ways that are not always visible. Air currents that mobilize toxics from distant smokestacks are hard to see; water currents that run under the surface are even harder to detect. Pristine, remote Lake Superior had seemed as if it couldn't easily be contaminated from industries located in other regions, but in the 1950s, scientists began to realize that assumption was wrong.

The most important assumption was that if something is very dilute in the water, it would be dilute in the organisms or animals that lived in that water. After the war, however, research on radioactive isotopes began to complicate that model. Postwar nuclear testing had worried many people, but the government had issued comforting assurances that radioactive waste that fell to the earth would be diluted in such low amounts that it would be harmless. Because contamination reports were initially classified, scientists could not access data to challenge this reasoning. As Barry Commoner said in an interview with *Scientific American:* "World War II had hardly ended when . . . the U.S. and the Soviet Union began testing new and nastier [weapons], creating enormous amounts of radioactivity that spread through the air worldwide, descending as fallout. . . . The tests were done in secret, marked only by Atomic Energy Commission [AEC] announcements that the emitted radiation was confined to the test area and, in any case, 'harmless.' This convenient conclusion reflected the AEC's assumption that the radioactive debris would remain aloft in the stratosphere for years, allowing time for much of the radioactivity to decay."[17]

In 1954, some government reports about nuclear testing were declassified, allowing scientists to examine records from weapons tests. Biologists, geneticists, ecologists, pathologists, and meteorologists began to explore the ways that radiation affected marine organisms, particularly the phytoplankton and zooplankton at the base of aquatic food chains. They asked: If radioactive isotopes were taken up by marine algae, what would happen to the fish that ate that algae? Movement of tiny amounts of contamination became core to the inquiry, illuminating concepts that had existed in the chemical literature but had not been widely applied to concerns about biological pollution: bioaccumulation and biomagnification. Although the terms are often used interchangeably in the popular literature, they are subtly different. Bioaccumulation happens within individuals when a substance is absorbed by an individual faster than it is metabolized, broken down, and excreted. Therefore, it refers to the increase in the concentration of a substance within a single organism's body. Biomagnification occurs across trophic levels, when substances become concentrated as they move up the food chain.[18]

Wildlife biologists began to write of their concern soon after the 1954 release of nuclear test results. In 1956, T. R. Rice of the Fish and Wildlife Service published a study showing that marine algae concentrated radioactive materials, writing, "Some elements are taken up in such large quantities that they are concentrated considerably over their amounts in sea water . . . [and] result in a rapid concentration of the activity in the planktonic algae. Consequently, animals feeding upon the algae could be expected soon to incorporate into their bodies a large percentage of the radioactivity." In 1958, J. J. Davis and R. F. Foster, two scientists from Hanford Atomic Labs, examined the ways that radioactive fallout could affect entire food chains. They warned that "where biological systems are involved, the organisms have accumulated certain isotopes to many times the initial concentrations in the water." Two years later, another report from Hanford Labs showed that "radioactive debris from distant nuclear weapons tests resulted in a tenfold increase" in rabbit thyroid glands. Fetuses and young rabbits "consistently contained greater concentrations of fission products than adult animals." Strikingly, greater distance from the release sites did not reduce risk: "Fission product concentrations in respective tissues of all kinds of animals sampled were essentially equal, irrespective of distance from Hanford process stacks."[19]

In 1960, when Rachel Carson was writing *Silent Spring*, she wrote a foreword for a revised edition of *The Sea Around Us* that incorporated the new research about biomagnification, noting, "Tuna over an area of a million square miles surrounding the Bikini bomb test developed a degree of radioactivity enormously higher than that of the sea water." In *Silent Spring*, she went into much more detail about bioaccumulation and biomagnification, arguing that the concepts applied to pesticides as well as nuclear fallout.[20]

Northern communities illuminated a central paradox for Carson: remote, seemingly pristine ecosystems were intimately tied to an industrial-

izing, militarizing world. In 1963, in a talk just before her death, Carson
spoke of lichens in the far north that received nutrients directly from the air
and so "pick up large amounts of the radioactive debris of fallout." She discussed the fact that "Cesium^{137} also travels through this arctic food chain,
to build up high values in human bodies." Arctic Indigenous peoples concentrated these toxics and "were carrying heavy body burdens of Ce^{137}."
Above the Arctic Circle, Carson reported, seven hundred Indigenous
peoples living in four different villages had Ce^{137} levels that "were about 3
to 80 times the burden in individuals who had been tested at Hanford."[21]
Phytoplankton, Carson wrote, can take up persistent contaminants, such
as DDT, from the water. Small fish and zooplankton eat huge quantities
of that phytoplankton, concentrating the toxic chemicals in their bodies.
Each step in the food chain intensifies this process, until predatory fish and
marine mammals have concentrations of chemicals that can be millions of
times greater than the concentration in the water—and the northern people who ate those animals were at the highest risk, even though they were
distant from the initial releases. These ideas shattered the dilution and assimilative capacity models that formed the basis for pollution control.

Canada: Toxic Exuberance

Even as scientists were developing new understandings of pollution bioaccumulation and mobility, the pulp and paper industry along Lake Superior's Canadian shore was booming, placing intense demands on the forests. Between 1930 and 1980, to meet the growing demands for wood, Great
Lakes Paper more than doubled its timber limits, expanding from just under 10,000 square miles to 21,000 square miles. Cords of wood harvested
grew even faster (figure 3.4). As a Great Lakes Paper internal document
notes, "The three decades following the Second World War were years of
maturity and phenomenal growth for The Great Lakes Paper Company."
By 1950 the United States was purchasing six million tons of Canadian
newsprint—nearly twice the 1945 quantity.[22]

To produce wood for the expanded mills, foresters turned to the
new organochlorine pesticides. Canadian provincial agencies encour-

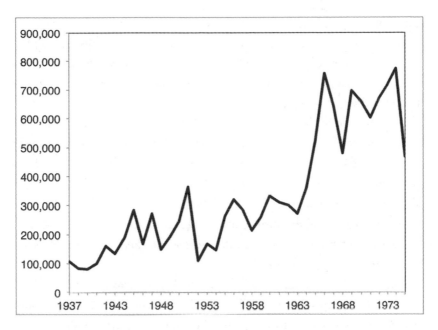

Figure 3.4 Cords of wood harvested by Great Lakes Paper between 1937 and 1975. As Great Lakes Paper grew in the decades following World War II, the company harvested substantially more forest. (Data from Great Lakes Paper Company records, Thunder Bay Historical Society.)

aged a massive increase in pesticide use, particularly the persistent organic pollutants. The postwar boom years in Canadian paper rested on the faith that low-value boreal forests could be transformed into productive timber factories with the help of miracle chemicals, particularly DDT. Insect attacks on boreal forests have always been part of their disturbance ecology. Eastern spruce budworm moths lay their eggs during the summer on conifer needles, particularly balsam fir and white spruce. Caterpillars overwinter and in the spring feed upon the trees, killing the trees if enough budworm are present. Large-scale infestations of the eastern boreal forest occur in cycles of roughly thirty-five years. The timing and extent of those infestations depend on spring weather (the caterpillars particularly like warm dry springs) and host populations (large expanses of mature balsam fir provide an excellent food source for an exploding caterpillar population). While these insect cycles are natural, after World War II, they began to complicate commercial efforts to fully exploit the boreal forest for pulp and paper production. Dead trees and massive fires signaled to foresters an unhealthy forest that required chemical interventions to achieve full productivity.[23]

When spruce budworm populations exploded in the late 1940s, foresters on both sides of the border were armed with new technologies made possible by the war: DDT that could be sprayed over millions of acres from planes released from military service. Aerial spraying of DDT did indeed suppress budworm populations, but only temporarily. By killing off 95 percent to 98 percent of the spruce budworm in an area, DDT spraying kept the budworms from killing off all the local spruce and fir. But that meant a superabundant food source for the 2 to 5 percent of budworms that had managed to survive each pesticide application. Budworm epidemics had, historically, collapsed quickly when budworms killed off their food supply. But now DDT actually prolonged the budworm cycles, leading to ever more defoliation and ever more spraying of DDT in an attempt to control the outbreaks. As the botanist George Woodwell noted: "Spraying half a pound of DDT in oil per acre could reduce that year's budworm population by 95 to 98%, but next year the remaining population would explode. Spraying only prolonged the outbreak, in fact, because while it kept the trees from dying, that meant the few remaining insects had unlimited food, and their populations could explode."[24]

Foresters tried to manage the boreal forests by removing small-scale natural disturbances, but in so doing, they likely increased the intensity and frequency of large-scale disturbances. Clear-cutting, replanting with white spruce and other species susceptible to budworm, fire suppression, and the use of pesticides all intensified budworm outbreaks. For example, the infestation of 1910–1920 defoliated 25 million acres. The infestation of 1945–1955, when DDT was first used heavily, defoliated more than twice the earlier infestation: 62 million acres. And the infestation of 1968–1985 defoliated even more: 136 million acres. As a comparison, the combined area of New York, Pennsylvania, Maryland, West Virginia, Virginia, and North Carolina is about 141 million acres.[25]

Spraying DDT did not stop the budworm, but it did ignite concerns about the environmental effects of massive spray campaigns in the boreal forests. In *Silent Spring,* Rachel Carson wrote of the "rivers of death" created by the intense DDT spraying in the boreal forests of Canada.[26] After *Silent Spring* was published in 1962, another five years would pass before aerial spraying of DDT ended in New Brunswick. In those five years, 12.5 million pounds of DDT were sprayed each year over the boreal forests of that one province alone. Not until 1985 did the Canadian government completely ban the use of DDT in forestry (although existing stocks could be used until 1990).

Global Transport of Pollutants

Wildlife within the boreal forest was clearly affected by DDT, but so were creatures living continents away, changing scientific understandings of toxic mobilization. In the 1950s, George Woodwell was a young botany professor at the University of Maine when the forests he was studying in northern Maine were doused with DDT. Woodwell grew concerned when his investigations showed that only half the DDT sprayed from the planes actually landed in the forests below. The rest seemed to vanish, and Woodwell set out to figure out where it went. He learned that the DDT solution dried into tiny crystals that could be easily dispersed on air currents and eventually be deposited tens of thousands of miles away. DDT residues, Woodwell learned, were appearing not just in boreal lakes but also in the tissues of seals as far away as Antarctica.[27]

Global transport pathways have existed "as long as the Earth has had oceans and an atmosphere," as the Canadian scientist Robie Macdonald writes. World War II, however, intensified the release of persistent contaminants, leading to widespread atmospheric contamination of the north. These transport processes are complex and difficult to predict, complicating efforts to regulate safe levels of contaminant release from industries. Airborne contaminants may take just days or weeks to move to the most remote locations. But if they make their way into water currents, they may take years to get very far. Contaminants that become airborne don't stay up there forever. If they bounce against buildings, they may stick to them. If a heavy rainfall comes along, contaminant bound to particulates in the air may come down. Some contaminants may remain hidden under the water or snow, captured in sediments for long times. But when disturbed, they may bounce back into the atmosphere, moving northward.[28]

Organochlorines such as DDT and toxaphene are chemically persistent and semi-volatile. They have vapor pressures high enough so that they readily evaporate, cycling between gaseous and condensed phases in the environment. The more chlorine in them, the more persistent they are. Some toxics have "one-hop" pathways: they bounce into the atmosphere once, and after they come back down, they stay down. But many contaminants bounce up and down, via the so-called grasshopper effect, reentering the atmosphere after initial deposition.[29]

Cold northern places have a pronounced vulnerability to contamination, as Carson argued. Contaminants persist for a long time in cold water without being broken down or escaping through evaporation. Indigenous peoples of the north often depend on "country foods"—high-fat animals at the top of their food chains that bioaccumulate extremely high levels of contaminants.[30]

Mercury was released in significant quantities by the pulp and paper industry as part of the chemical bleaching process. Like toxaphene, mercury acts as a grasshopper, bouncing northward. Unlike toxaphene, it is a natural chemical that has been part of global biophysical cycles, existing long before people showed up on the scene. But though it is natural, industrial activity has caused a "two- to fivefold increase in its concentration in air and marine surface waters of the northern hemisphere since the pre-industrial era." Fossil fuels contribute over half the global load of mercury—estimated at about 2,464 tons of mercury released in 1995 alone. China was the single largest emitter of mercury from fossil fuel consumption, with 546 tons released in 1995. Much of that makes its way on global atmospheric currents into the north, particularly the Bering Sea, Alaska, and the western Arctic.[31]

After the release of elemental mercury into the atmosphere, some portion is degraded, but much of the mercury is deposited via precipitation

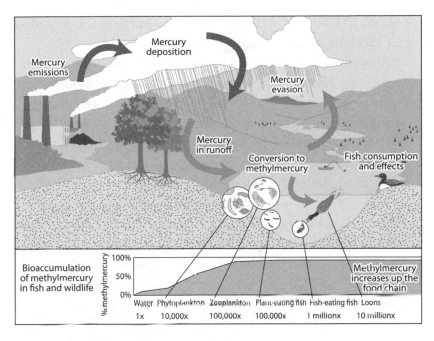

Figure 3.5 Mercury methylation and deposition cycles. (Drawn by Bill Nelson, modified from Great Lakes Mercury Connections 2011 Report BRI 2011-18.)

into soil and water (figure 3.5) The elemental mercury is not particularly toxic to fish or people, but when certain bacteria are present in acidic conditions, elemental mercury is converted to methylmercury, which is quite toxic to neurological systems. Additionally, methylmercury biomagnifies up food webs, until concentrations in predatorial fish can be millions of times greater than concentrations in water.[32]

Historical records from northern and Arctic lake sediments show a continuing increase in mercury deposition through the twentieth century. This mercury doesn't come directly from coal burning, because few coal-burning power plants are clustered in the Arctic. Rather, Arctic levels of mercury are so high because the north receives airborne mercury from China and the lower Great Lakes. Additionally, changing climate processes may be increasing the transfer of mercury from soil into biota.[33] Arctic soils contain an archive of contaminant mercury accumulated since the industrial revolution, and this mercury may be released as permafrost thaws, fires burn, and floods inundate northern wetlands.

Dioxins are another persistent, bioaccumulative chemical once released into the Lake Superior watershed by the paper industry. When the chlorine in the kraft pulp-bleaching processes combines with the organic matter in pulp, it forms chlorinated organic compounds known as dioxins. Dioxins cause birth defects and cancer, accumulate up the food chain in fats, and create hormonal havoc. Manufacturers and workers have recognized their risks since the 1930s, but that information was not provided to the regulatory authorities during the expansion of chemical bleaching in the paper industry. Barry Commoner wrote in 1995, "The history of dioxin is a sordid story—of devastating sickness inflicted unawares, on chemical workers; of callous disregard for the impact of toxic wastes on the public; of denial after denial by the chemical industry; of the industry's repeated efforts to hide the facts about dioxin and, when these become known, to distort them." In April 1930 African American workers in the Anniston, Alabama, Monsanto factory who worked with chlorinated diphenyls (the early term for PCBs) became very sick. The process that produced PCBs also produced dioxin-like substances as a by-product. The workers' illness was dismissed as "the usual temperament of the Negro toward work" in a 1936 case study published in the *Archives of Dermatology and Syphilology* by two Atlanta physicians, who reported severe chloracne, lassitude, loss of appetite, and reproductive problems such as loss of libido.[34]

The next key warning about the toxicity of dioxins emerged in 1957, when chick-edema disease killed millions of young chickens in the eastern and Midwestern United States. Fat scraped from tanned hides had been added to chicken feed, and those hides had been treated with a "fleshing grease" contaminated with dioxins. The distillation process that was supposed to make this waste product safe for chicken feed actually concentrated the dioxin impurities.[35]

Even after the Times Beach debacle in the early 1970s, when dioxins contaminated dozens of sites across Missouri and forced the abandonment of communities, the widespread use of materials containing dioxins continued, spreading the chemicals throughout North America. Soon after Times

Beach broke into national news, researchers found deformities in wild birds around the Great Lakes that looked suspiciously similar to chick-edema disease. Many eggs failed to hatch, and many chicks that managed to hatch had twisted beaks, crooked legs, deformed claws and poorly formed feathers. "I could only describe the situation as catastrophic," said researcher Michael Gilbertson. In 1973, Gilbertson noticed that many of the wild birds' symptoms were quite similar to those in chickens with chick-edema disease. In 1991, he published a key paper with Glen Fox, James Ludwig, and Timothy Kubiak that offered evidence that exposure to dioxins and certain PCBs were the likely cause of the outbreaks.[36]

Paper production indirectly increased dioxin exposure through herbicide spraying of forests. The popular forestry herbicide 2,4,5-T was contaminated with dioxins during its manufacture. Agent Orange, used as an herbicide in Vietnam, contained 2,4,5-T mixed with 2,4-D, which meant that it contained significant levels of dioxins. American soldiers and Vietnamese civilians were exposed, leading to international condemnation. But while anger bubbled up over the contamination of soldiers and civilians in Vietnam, use of 2,4,5-T boomed in forestry, part of the effort to rationalize and intensify forest production. Despite the military ban on Agent Orange and a warning from the U.S. attorney general that 2,4,5-T contaminants "may present an imminent hazard to women of child-bearing age," domestic uses of 2,4,5-T grew, even as understandings of dioxin's toxicity grew. Rather than ban 2,4,5-T because of its dioxin content, the young Environmental Protection Agency (EPA) "yielded to manufacturers' repeated calls for further study." By 1975, scientists from EPA's laboratories were finding "frightening levels of dioxin in tissues of animals and birds from sprayed roadsides, in beef cattle grazed on sprayed range land, and in human mothers' milk from sprayed areas of Oregon, Texas and Vietnam."[37] Under pressure from industry, the EPA took no action until a group of rural Oregon residents filed a lawsuit and won a federal court decision banning the use of dioxin-contaminated herbicides on national forests.

On December 2, 1980, the Canadian government released a report showing high levels of dioxin in gull eggs and tissue from Saginaw Bay and other areas of the Great Lakes region. Internally, EPA scientists noted that the dioxin levels in Great Lakes fish were high enough to cause a significant increase in cancer rates among consumers. But rather than issue an advisory and quarantine the contaminated fish, the Food and Drug Administration agreed with industry that "there is considerable uncertainty about dioxin's effects on humans," so setting an action level would be "premature." An internal EPA draft report from Region 5 disagreed with this conclusion, instead arguing that dioxin levels found in Great Lakes fish presented a "grave cancer hazard to consumers." When the internal EPA report was leaked to Toronto's *Globe and Mail,* international fury resulted.[38]

In 1985, the EPA released a cancer risk assessment that identified dioxins as the most potent synthetic carcinogenic chemical known, setting the acceptable level of contamination at 1 part per billion (ppb). This angered the forest products industry because chemical pulp bleaching was releasing significantly more dioxin than that in rivers. Rather than agreeing to institute effluent controls or change bleaching chemicals, the pulp and paper industry chose to discredit the science, delaying regulation of dioxins in the pulp industry.[39]

Paper production created a waste product called sludge that was heavily contaminated with dioxins. Paper companies in the Lake Superior basin disposed of sludge either by having it spread on farm fields or by sending it to municipal sewage facilities to be landfilled or burned. All three possibilities ensured the transport of dioxins into surface water, groundwater, or the atmosphere. For example, the Lake Superior Sanitary District in Duluth accepted waste sludges for incineration from the Potlatch Corporation pulp mill in nearby Cloquet, Minnesota, releasing dioxins into the air. By 1985, fish samples collected downstream from pulp and paper mills in Wisconsin were consistently contaminated with dioxins. Carp downstream from pulp and paper mills on the Wisconsin River had particularly high levels of dioxin. When the EPA received these data from the state, the EPA concluded that the chlorine-bleaching process in kraft-process mills was a "potential" dioxin source, but the agency continued to delay regulation.[40]

In 1987, Greenpeace published a report about dioxin contamination in the pulp industry that claimed: "There is enough evidence to be certain that chlorinated dioxins are an unwanted byproduct of all pulp and paper mill production processes using chlorine. Just as certainly, there are emergency steps that should be taken to vastly reduce the levels of dioxin emissions in the industry. . . . Instead of taking such steps, industry and federal government officials have conspired to conceal the problem." Although the report gathered international attention, three years after its release, over 90 percent of kraft pulp plants in the Great Lakes basin were still releasing dioxins into the water. Even after dioxins from milk cartons were found to be leaching into the milk, Red Caveny, president of the trade group American Paper Industry, insisted in an interview with the *New York Times* on May 1, 1990, that "the industry was already doing most of the things that would be required by formal regulation." While James Benson, acting commissioner of food and drugs, said dioxin was "not a major health problem," the EPA eventually agreed to force the paper industry to stop dumping dioxin-containing wastes into streams, an action that took effect in 1995. [41]

Boreal Pulp Forestry and the Health of Communities

The postwar boom in pulp and paper production around Lake Superior had rippling effects on the health of forests, watersheds, and human communities. Long before the northwestern Ontario paper boom, First Nations communities had been actively managing the boreal forests with fire management, wild rice and medicinal plant cultivation, and hunting. First Nations communities were displaced from boreal forests where logging operations were planned. Even though they retained hunting and gathering rights on ceded territories, forest operations rendered those rights almost impossible to exercise. [42]

In the 1987 environmental assessment process for forestry in northern Ontario, members of the Grand Council (Cree) gave testimony about the harm they had suffered from forestry. They recounted that "cutting of large tracts of timber affected wild plants, game, and fur-bearing animals" and "traditional occupations involving off-reserve forest resources, such as hunting, fishing, gathering, and craft production, were not protected from the detrimental effects of non-Indian timber harvesting." Tribal lawyers stated: "Ojibway use of non-reserve forests, for shelter, food, industry and

crafts, medicine and communication and arts, was affected by forest industry harvests in particular cutting areas, by the creation and management of
dams and hydroelectric facilities, and by changing species composition of
the forest. . . . We have heard Ojibway Elders speak of environmental effects
of particular cutting practices, particularly clear-cutting, on fur harvests.
We have also heard, again from Ojibway Elders, comments regarding lack
of availability of certain tree and plant species, such as birch, cedar and
maple." White-tailed deer expanded their range in response to logging, exposing moose to parasitic flatworms. Woodland caribou declined as well
after logging.[43]

Pulp wastes were particularly devastating to tribal fisheries. Wastes
from the Fort Frances, Ontario, pulp mill were pumped into the river from
1905 to 1910, polluting sturgeon spawning grounds and "causing a decrease
in sturgeon population from which it has never fully recovered." In 1913,
members of the Little Fork Reserve petitioned the premier of Canada, noting that the dams for pulp mills had destroyed their livelihoods by killing
fish. When the Ontario-Minnesota Pulp and Paper Company (Kenora Division) put a dam at Shoal Lake to bring pulpwood through to Lake of the
Woods in 1942, the acting Indian agent for the Shoal Lake Indians wrote
that hay lands and wild rice beds had "all been flooded out."[44]

Indigenous women in Canada exposed to DDT spraying during the
spruce budworm campaigns may have experienced reproductive harm
from the endocrine-disrupting chemicals involved. Cree women who reside in Canada's boreal forests currently experience high rates of prenatal,
stillbirth, and newborn deaths—2 to 2.5 times the national average. Persistent organic pollutants such as DDT were used widely in boreal forests,
and growing evidence shows that boreal forest ecosystems may intercept
and retain these compounds. Predatory fish that reside in boreal aquatic
ecosystems bioaccumulate the chemical, and for people who eat those fish,
human exposures were significant.[45]

In 2001, Matthew Longnecker and his colleagues at the National Institute of Environmental Health Sciences measured levels of dichlorodi-
phenyldichloroethylene (DDE)—a chemical formed when DDT breaks
down—in the stored blood sera of 2,380 mothers who gave birth between

1959 and 1963, when DDT was being heavily used in agriculture and forestry. Of these women, 361 delivered prematurely. The higher the level of the endocrine disruptor DDE in the mother's blood, the greater the odds of preterm birth.[46] These associations are not proof that boreal forest spray campaigns caused the elevated rates of reproductive problems experienced by Cree women. They do suggest, however, a plausible link between exposure to persistent organic pollutants and fetal harm.

Grassy Narrows was one of the worst poisoning episodes associated with pulp and paper production. In 1962, Dryden Chemical Company opened a plant that produced paper-bleaching chemicals, using mercury cells in a chlor-alkali process. These chemicals were destined for the nearby Dryden Pulp and Paper Company. Both companies were subsidiaries of the giant British multinational company Reed International (now the academic publisher Reed Elsevier). Dryden Chemical Company discharged its mercury-laden effluent directly into the Wabigoon–English River system, dumping an estimated 22,000 pounds of mercury over the next decades. The mercury contaminated fish in the river, making them poison for people who ate them.[47] These fisheries had been the basis of the subsistence and the economy of three Indigenous communities: Asubpeeschoseewagong (Grassy Narrows), Wabaseemoong (White Dog), and some members of Wabauskang. More than half a century later, mercury levels in fish are still elevated, and members of the Indigenous communities are still struggling with the health and cultural impacts.

In 1970 the government ordered Reed International to stop dumping mercury into the water system and closed the local fishery, while allowing airborne release of mercury until 1975. Closing the commercial and subsistence fishery devastated the economy of the Grassy Narrows Ojibway. Not only did they lose their major food source, the summer fishing resorts that had brought in a recreational fishery and sustained the community's economy also shut down. Mercury poisoning, known as Minamata disease (named after the Japanese town where widespread mercury poisoning was caused by eating contaminated fish), has been identified.[48] Mercury was well known to be poisonous when Dryden began dumping it into local waters, and Minamata disease had already been discovered in Japan.

The historic legacies of pulp mill pollution still haunt local communities that are faced with toxic burdens as well as the economic burdens of cleanup. Changes in pulp and paper markets have meant that it is no longer cost effective to run most of the huge pulp mills, even though modernization has reduced employment and labor costs. Increased energy costs combined with decreased prices for pulp have made Canadian pulp less competitive with pulp produced in the southern hemisphere, where trees grow more quickly and labor costs far less. Yet although the jobs have left, the toxic legacies persist, with contaminated Areas of Concern lining the former paper mill communities along Lake Superior.

Taconite and the Fight over Reserve Mining Company

IN 1947, JUST AFTER WORLD WAR II, depletion of iron ore from the Lake Superior Iron District seemed to be a looming national security risk. Politicians worried about the so-called iron ore dilemma in the United States—the fear that dwindling supplies of high-grade iron ore in the Lake Superior District threatened Cold War strategic interests.

The iron ranges of Minnesota, Wisconsin, and Michigan, collectively known as the Lake Superior District, had "nourished the U.S. economy for half a century, through two World Wars."[1] By the end of World War II, the district provided 85 percent of the United States' supply of iron ore (figure 4.1). But increased wartime steel production meant enormous pressures on dwindling iron ore reserves. By 1945, much of the Lake Superior region's accessible high value iron ore had been extracted.

Iron was a key component of steel, and steel was essential for industrial and military purposes. As the following case study of Reserve Mining Company along Minnesota's north shore argues, postwar concerns about iron depletion led American mining interests to promote technologies and tax incentives to exploit taconite ore bodies. Taconite is a lower-grade iron ore containing about 25 to 30 percent iron content (typically in the form of magnetite, a magnetic iron mineral) mixed with chert (composed of fine-grained silica). Though taconite was formerly considered a waste product, after World War II companies became interested in its potential. Yet this low-grade material required expensive new processing technologies to be profitable, while creating new environmental consequences, particularly concerning finely ground tailings and the use of water.

Figure 4.1 One end of the Hull-Rust-Mahoning pit near Hibbing, Minnesota, the largest open-pit iron mine in the world when this photo was taken in 1941. (John Vachon, photographer, Farm Security Administration/Office of War Information. Library of Congress LC-DIG-fsa-8c20455.)

This chapter and the two that follow explore the growing controversies over pollution from taconite. As taconite iron ore mining boomed
in the Lake Superior basin in the three decades after World War II, faith
in cooperative pragmatism began to clash with new industrial developments and new understandings of pollution mobility. The debates in 1947
over where tailings should be dumped and the ways pollution might move
through broader spaces are the subject of this chapter. The ways that taconite pollution became core to a growing environmental movement in the
1960s and 1970s is explored in chapter 5. And twenty-first-century controversies over a proposed new taconite mine that would have been sited
just upstream of the Bad River Band's reservation on Lake Superior are
analyzed in chapter 6.

In 1947 Reserve Mining Company proposed to mine taconite at the
Peter Mitchell Mine in the eastern end of the Mesabi Range, then construct
a railroad to transport the ore to a new taconite-processing facility at Silver
Bay on the shores of Lake Superior, about fifty miles northeast of Duluth
(figure 4.2). The company would dump its tailings directly into Lake Superior near Silver Bay, saving enormous sums on land disposal. To make the
processing of taconite feasible, Reserve Mining required over 500,000 gallons of water per minute from Lake Superior, and it would need to dump
about 67,000 tons of tailings each day into the lake — eventually totaling 400
million tons of tailings. Fearing significant environmental consequences,
the state of Minnesota held an extensive series of hearings in 1947 to decide
whether taconite mining, processing, and tailings disposal would be safe
for Lake Superior. Thousands of pages of hearing testimony and supporting materials were opened to the public in 2013, and these archives reveal
that regulators, companies, and communities took possible environmental
threats quite seriously, just as they took the potential benefits to national
security and the regional economy quite seriously.

To understand the intensity of these debates over pollution, and the
ways that the changing models of pollution discussed in the previous chapter influenced those debates, we need to step back a moment to look at
the importance of the region's iron ore for national security and economic
development. The iron mines of the Lake Superior District had been key

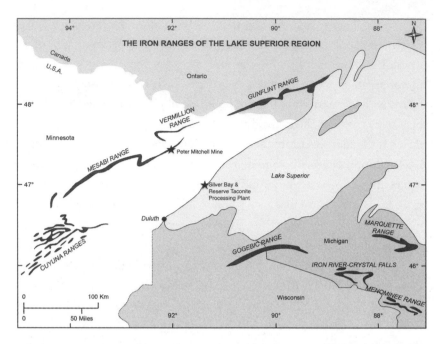

Figure 4.2 Locations of the Reserve Mining Company taconite-processing plant near Silver Bay, Minnesota, and the Peter Mitchell taconite mine in the Mesabi Range that supplied the Silver Bay plant. The Mesabi Range was the most productive of all the iron ore ranges in the Lake Superior District. (Drawn by Bill Nelson, modified from W. F. Cannon, U.S. Department of the Interior USGS Report, via Wikimedia Commons.)

for late nineteenth-century American industrialization. As the main ingredient in manufactured iron and steel, iron ore was essential for building automobiles, railroads, skyscrapers, barbed wire, tanks, bombs, and bullets. The opening of the Soo Canal at the bottom of Lake Superior in 1855 had spurred iron mining in the Lake Superior District, for the canal reduced the cost of shipping ore to markets. Before the canal, rapids along the St. Marys River connecting Lake Superior and Lake Michigan had presented a significant obstacle. Any iron ore headed for the lower Great Lakes had to be unloaded from the ship, hauled around the rapids, and loaded on another boat. Federal funds helped to create a shipping and railroad infrastructure that overcame nature's obstacles, enabling large-scale iron mining

Figure 4.3 Iron ore boat *D. G. Kerr* loading at the ore docks, Duluth, about 1930. (Photo HD3.15 P54,12381, Minnesota Historical Society.)

to develop. Duluth eventually developed into a major port serving the iron industry (and eventually the Great Plains wheat industry as well). Investors were excited. Boosters promised that remote Lake Superior villages— Duluth, Marquette, Ashland—would soon rival Chicago (figure 4.3). The Soo Canal stimulated the flow of minerals out and the flow of capital in, but most of the profits moved out of the region along with the iron ore.[2]

Between 1896 and 1900, the American steel industry had experienced a radical transformation. Small companies were replaced by large corporations that controlled not just steel mills, but also the iron mines that supplied those mills. By 1900, nearly three-quarters of all Lake Superior iron ore came not from mines operated by smaller companies, but rather from mines that were "either owned by or under long-term lease" to the largest American steel companies: Carnegie Steel, Federal Steel,

National Steel, and several others—all to be absorbed into U.S. Steel in 1905. Engineer Edward Davis wrote in 1942 that "79 per cent of the remaining open pit direct ore is owned or controlled by the United States Steel Corporation."[3]

Iron ore in the Lake Superior basin fell into two broad types: high-grade ores (often called "direct shipping ores" because such ores required minimal processing before shipping) and low-grade ores, which required beneficiation, an expensive and technically challenging process involving crushing, screening, grinding, magnetic separation, filtering, and finally drying.[4] Before World War I, most iron mining in the Lake Superior Basin focused on higher-grade hematite, which contained such high concentrations of iron (50 to 70 percent) that it did not have to be processed before shipping. Some of these mines, such as those in the Gogebic-Penokee Range, were deep-shaft mines, and others were open-pit mines. But in both cases, although surface disturbances differed, the high iron content meant that waste was minimal. In particular, no fine tailings were produced near the mining site, and no water was contaminated or consumed in processing.

Before World War I, some companies began developing lower-grade ores called "washable ores" containing about 30 to 45 percent iron. This lower concentration meant that washable ores needed beneficiation (processing) near the mine to remove the silica and concentrate the iron before shipping it to steel manufacturing plants.[5] Waste products from processing called tailings were deposited in inland lakes throughout the Minnesota iron ranges, and a great deal of water was diverted to serve the beneficiation process. Conservationists and locals concerned about the pollution of inland lakes were vocal in their protests, encouraging mining companies to look for tailings disposal options that didn't fill inland lakes.[6]

In 1947, Reserve Mining Company—under the ownership of Armco Steel Corporation and Republic Steel Corporation—applied for permits to process taconite on the shores of Lake Superior near the small community of Silver Bay. Lake Superior would supply both the abundant water needed for taconite processing and a convenient location for tailings disposal. The company proposed to construct a launder—a long, sloped concrete channel allowing gravity to move a slurry of waste material and water—to dump about 67,000 tons of tailings each day out into the lake (figure 4.4). Eventually 400 million tons of tailings would be dumped in the lake. Reserve promised that once the mine reached full capacity, it would produce more than 1 percent of the country's total iron production, becoming an economic engine for an impoverished region.[7]

Taconite is a low-value ore, with only 20 to 30 percent iron content. Unlike the washable ores, it was found in extremely hard formations, banded with quartz and other hard rocks, and the material it was mixed with was extremely difficult to remove. The waste material couldn't just be washed away after grinding the ore. Mining taconite required extensive technological and financial investments in beneficiation, an energy- and water-intensive series of steps.

With the development of taconite, mining for iron was to become less a simple matter of extracting valuable ore from the ground with minimal processing and more a case of manufacturing production. As Reserve Mining Company officers told the hearing commission in 1947, "The entire process of mining, crushing, fine grinding and sintering or agglomerating [taconite] is an extremely complicated and costly process. . . . Essentially it is a manufacturing industry rather than mining; the manufacture of a useable iron ore from materials which are absolutely worthless in the ground."[8] Timothy LeCain argues in *Mass Destruction* that the development of low-grade, open-pit mining involved two key elements: a rationalized, envirotechnical system and far greater environmental destruction. LeCain describes the very visible environmental deterioration that accompanied copper open-pit mines in the American West, including dead trees from the sulfur and belching clouds of poisonous gases contributing to lung diseases.[9] Industrialized mining became dominated by large corporations using enormous machines to extract low-grade deposits from open pits. In the Lake Superior basin, the massive expansion in taconite technologies borrowed directly from copper technologies LeCain describes in *Mass Destruction*.

Figure 4.4 Taconite launder. Conveyor discharges taconite tailing residue into Lake Superior at Reserve Mining Company's taconite plant in Silver Bay, June 1973. (Donald Emmerich, photographer, U.S. National Archives and Records Administration record 3045077.)

Historian Jeffrey Manuel notes that the engineer Edward Davis, who spent much of his career promoting taconite, "did not invent the technologies used in taconite milling from scratch." Rather, he "drew from technical developments in ore concentrating and milling that had revolutionized the mining and minerals industry since the mid-nineteenth century." Davis understood that taconite was more than just a technical challenge. It was also a political one. He lobbied the Minnesota legislature to pass legislation favorable to taconite processing. The legislature responded by reducing the taxes that taconite companies paid and changing property laws that had protected land owners from expanding mines. In 1941 Minnesota passed a law "virtually exempting taconite property from the local ad valorem (property) tax, a tax particularly objectionable to mining interests."[10] Four years later, the legislature passed a law allowing eminent domain to be used to seize private property for tailings waste piles. These decisions shifted many of the environmental and economic costs of mining to communities while channeling profits to industry. They did what engineering alone could not do: allow taconite production to boom. After the 1941 and 1945 laws, companies quickly moved to create the infrastructure that would enable them to tap the vast taconite reserves along Minnesota's north shore.

Mining industrial developments were core components of envirotechnical developments across the north—developments in which engineers mobilized state power (and state power mobilized engineers) to create what James Scott calls a high-modernist vision of legibility and order. Like other envirotechnical systems in the post–World War II north, the expansion into taconite mining receive significant state support, required enormous sums of capital, and irrevocably transformed freshwater ecosystems and the communities that depended on them. At the 1947 permit hearings for Reserve Mining Company's proposed project, the hearing chair called taconite "a pioneering adventure." Engineer Edward Davis chose that word as well for the title of his book about his adventures in mining: *Pioneering with Taconite*.[11] A close examination of the permit hearing records that were made available to the public in 2013 helps us understand the underside of that pioneering adventure: the effects on the communities and ecosystems.

Taconite Hearings, 1947

In 1947, the Minnesota Department of Conservation and the Water Pollution Control Commission held a series of public hearings on Reserve Mining Company's permits for their taconite processing plant on Lake Superior. The company requested permission to take half a million gallons of water from the lake every minute of every day and in turn to dump 67,000 tons of tailings into the lake each day.[12] Waste from a project this size dwarfed the waste from direct shipping ores and washable ores.

Chester S. Wilson, commissioner of the Minnesota Department of Conservation, served as hearing chair. He was a classic exemplar of the cooperative pragmatist tradition in the Great Lakes basin. A dedicated conservationist, he served as the Minnesota conservation commissioner between 1943 and 1954, conducting the permit hearings jointly between the Department of Conservation and the Water Pollution Control Commission.[13] An attorney by training rather than a scientist or engineer, by 1947 Wilson had become one of Minnesota's most important conservationists. He had helped develop key pieces of environmental legislation, and he served as legal counsel on a number of state and national environmental boards. Most important, he believed in the premises of cooperative pragmatism: namely, that pollution could be contained with technology, and that voluntary partnerships with industry were more effective than top-down regulations. He also believed that the state, not the federal government, should control conservation decisions.[14]

Before granting permits, Wilson held seven joint hearings to decide whether Reserve's plan to dump tailings into Lake Superior would be worth the risks to the watershed. The company spoke first, framing its request within the larger context of national security and the threat of iron depletion. William K. Montague, head lawyer for Reserve and former assistant attorney general of Minnesota, opened by saying: "By now everybody realizes that the future of the iron mining industry in the state depends upon whether or not the taconite industry can be developed. The state is in a position where it can see the rapid curtailment of mining operations and the early exhaustion of ores unless they can be supplemented and eventually replaced by the production of ore from taconite."[15] The engineer Davis spoke after Montague, essentially repeating his arguments about the importance of taconite for the state's economic development. Reserve's speakers expanded this argument to a national scale in the hearing, stating, "The development of large reserves in this area is recognized to be of national importance because a serious problem of national security is presented if the steel industry is dependent upon foreign sources for its iron ore."[16]

Iron and steel were certainly central to the nation's economic growth, but it is also important to recognize that depletion concerns were a political construct, revolving around security fears about what might happen if new sources weren't developed. Resource depletion is mediated by technology, rather than being an absolute measure of a quantity of a resource. If all you have is a spoon, an ore body will be depleted for you as soon as the loose surface ore is scooped up. But if you have the 5.5-million-pound explosives that taconite operations currently use, you can access the valuable iron from entire ore bodies, sources that were essentially invisible in earlier accounts of minerals because they were deemed forever uneconomical to mine.

Fears about iron ore depletion created a new discourse: we must relax environmental standards and tax policies to ensure American national security. The popular magazine *Fortune* warned that an iron ore dilemma was developing for North America. How long could the Lake Superior District compete with high-grade ores from such politically unstable countries as Brazil and Venezuela? And could the United States "risk the prospect of another war without a large and quickly available stock-pile of iron ore or pig iron?" The story, in other words, was not just about depletion of an important resource leading to pressures to develop new technologies. Rather, it was about the various ways different political and scientific actors manipulated concerns to drive exploration. As Manuel argues, depletion concerns in the Lake Superior basin were embedded in specific political contexts, intended to generate pressure for government funds and new tax policies that would benefit taconite mines over direct shipping ore mines.[17]

Yet even as proponents of the mine made the depletion and national security arguments, under questioning they admitted that dumping the tailings into Lake Superior was not a technological requirement. Rather, it was simply cheaper than inland disposal. Davis said, "There certainly are plenty of wasteland areas of Minnesota where tailings could be dumped without doing any particular damage," but Reserve "must save every penny in this operation in order to compete, in order to make this attractive, and if they can save a few cents per ton by using Lake Superior water, why that puts this state in a better position to compete with the ores from abroad and from New York state."[18] A few decades later, when pressure grew to close the facility, Reserve insisted that dumping the tailings in the lake was the only technologically feasible choice. Early testimony from the company, however, makes it clear that the company always considered alternatives.

Waste from the earlier washable ores set the stage for debates at the 1947 Reserve hearings over taconite wastes. In 1945, mining companies interested in developing a washable ore project called the Birch Lake and Dunka River Project had to figure out how to dispose of their tailings. Tailings in earlier washable ore projects had been deposited in inland lakes or in "wastelands" (wetlands). But these created nuisances because the "tailings are very fine and when they are dumped on the land they tend to make great dust storms in the area." The mining company behind the Birch Lake and Dunka River Project could not find willing sellers for the land they needed for their tailings pile because neighbors feared the dust storms and water pollution experienced on an earlier project. So the company appealed to the Iron Range Resources Commission, asking for permission to "resort to condemnation." The legislature agreed, allowing mining companies the right of eminent domain for land needed for waste disposal. Similarly, the legislature acted to allow corporations to use waters from the Superior National Forest in processing low-grade iron ore.[19]

In the 1947 hearings, conservation groups and local citizens alike were still infuriated by these 1945 laws that had allowed seizure of inland lakes for mining waste disposal. The Izaak Walton League spokesperson, Olin L. Kaupanger, testified, "We are opposed to granting any individual or corporation the right of eminent domain. . . . This act is the most damaging to natural resources that has ever been passed. The possibilities for destruction surpasses the greed of the early despoilers of this state, who confiscated at their pleasure."[20] One local worker agreed, protesting that corporations could now "take your property. . . . It imposes on the public. It gives a corporation a right they have no right to whatever."[21] Yet, ironically enough, the Izaak Walton League supported Reserve's permit request because members feared the seizure of more inland lakes if the project were blocked. Reserve argued that Lake Superior, unlike inland lakes, was large enough to dilute threats from mining wastes, and some conservation groups agreed with them.

William Montague, Reserve's lawyer, described the beneficiation process in terms that blurred the boundaries between natural and synthetic when he said, "The action of percolating water over long periods of time has at certain particular locations leached and softened the original iron formation, producing areas of partial enrichment." Beneficiation technology didn't replace these natural processes; rather, it intensified them. Montague continued: "The taconite industry as it develops will . . . try to *reproduce the processes of nature* in making this production" (emphasis added).[22] Mining proponents tried to reassure an uneasy public that their new technologies weren't some terrifying synthetic, pollution-causing blot on the natural landscape. Rather, they would simply improve on nature. During the permit hearings, this turn to nature became a way to reassure the public about the essential naturalness of the endeavor while simultaneously impressing the public with a sense of the project's grandeur.

H. S. Taylor, Reserve's head engineer, suggested that beneficiation was essentially a new form of alchemy. It could take "rock which is of no value whatsoever" and transform it with "the touch of the flow sheet." The engineer's symbol, the flow chart, could turn worthless rock into something less flashy than gold but even more useful for modern America: "ore suitable for blast furnace use, suitable for the steel mills, suitable to produce the material that makes America great."[23] Alchemy once promised the transformation of lead into gold to build the power of kings, enabling new empires of wealth and power. Beneficiation was a modern alchemy, harnessing the forces of nature and intensifying them with engineering. Yet the essential naturalness was still there, Reserve insisted.

The company stressed the brute force nature of its project at the hearings, reiterating rather than minimizing the massive quantities of water and waste involved. But the result, Reserve promised, would be pure and natural—and therefore nothing for the state's citizens to fear. Montague promised that Reserve would not use "chemicals of any kind. As a result, the tailings which go into the lake will have no chemicals added to them; will have no impurities of any kind added to them in the operation." Later he added that because the water would "not have any chemicals added," the waste would be purely natural, just "fine sand having the same chemical analysis approximately as the rocks along Lake Superior at the present time." The essential naturalness of the process became core to the company's arguments. Davis testified that beneficiation was similar to "the wave action and the action of the currents along the lake shore [that] is constantly grinding up these rocks and produces large amounts of silt that wash out into the lake." Just like natural sediments, Davis added, the mine tailings "will have no effect whatever on the lake. There will be no more effect on the lake than the wave action on Park Point has on the chemical analysis of the water."[24]

Reserve witnesses testified that because natural streams added silt to the lake, what could be wrong with Reserve contributing to that silt? Critics of the mine protested, pointing out that just because sediments were natural didn't mean adding more wastes was benign: "They [Reserve] speak continually about our streams, the amount of silt that is carried into the lake by our streams. I know that the streams carry a lot of silt into the lake . . . but is that any reason why we should double or treble it by man's help? It isn't."[25]

Chair Wilson asked the state to calculate how the Reserve tailings would compare in volume to silt deposited by the tributaries. John Moyle, the state's fisheries researcher, estimated that a year of Reserve's tailing would contribute about two-thirds of the silt added by natural processes, although he added this was "just a scientific guess."[26] (Eventually it emerged that Moyle's estimate was off by two orders of magnitude. In a single week, Reserve's tailings were the equivalent of a full year's worth of stream and river silt.) Reserve lawyers used Moyle's calculations to argue that if the plant would add less sediment than natural processes, surely no one could object.

Much of the debate in the hearings came to revolve around the question: What exactly is pollution? Can pollution be caused by something natural, or only by synthetic additives? Reserve insisted that because "no chemicals used are added to the tailings in the process, and no chemicals are added in that way," their tailings should not be classified as pollutants. Engineer Adolph Meyer testified that Reserve would add "no deleterious chemicals. It is only the finely ground rock which will go back into the lake so that there is no pollution involved in the project." Not everyone agreed. Mr. N. A. Nelson, a local who testified against the project, insisted that the tailings would make the water impure: "When this pollution gets into the water—I call it pollution—there must be some way to take that out of the water before it is fit to drink. Otherwise we will be drinking what is really harmful to humanity."[27]

Just as with the new persistent, bioaccumulative contaminants discussed in chapter 3, new forms of mobilization confounded regulation. Would tailings stay in place, affecting only the most local sites, or would they move underwater into new places, difficult to trace, track, and contain? Reserve insisted their tailings would stay within a deep trench near the launder (the concrete channel that dumped them onto the surface of the lake). Davis argued that in his laboratory experiments, the tailings sank to the bottom of the tank, and he insisted that this lab experiment meant Reserve's tailings would behave the same way, sinking into "deep valleys at the bottom of Lake Superior. There they would be out of sight forever and posterity would not have to cope with them."[28]

To reassure the commissioners, Davis brought a bottle of water with tailings in it to each hearing, and he shook the bottle at each hearing to show how tailings quickly settled at the bottom. Commissioner Wilson took these claims quite seriously, but numerous local witnesses objected to the bottle demonstration, pointing out that tailings in a bottle of water hardly represented the biggest lake in the world. Nelson spoke up again, arguing: "The currents, wave action, and what not is also going to carry it away, I would say 30, 40, and possibly a hundred miles, because the constant agitation will not let it settle as fast as it will in a bottle standing perfectly still. That is a very poor test and it shouldn't have been entered in here as an exhibit." A fisherman argued that bottles in court were one thing, but "the minute a storm comes—we know that people who live on Lake Superior know that storms are going to pick it [the tailings] up and don't know where it's going to end." The lake is more complex than a laboratory model, locals insisted, and the first storm will push the tailings out of local spaces. A critic of the mine got an engineer from Reserve to admit that "after the tailings leave your plant you have no control over them, is that correct?" Reserve engineer Taylor agreed: "No, we have no control over them."[29]

Sanitary engineers from Duluth were troubled by this lack of control, fearing that tailings would migrate southwest toward Duluth and enter their water supply, which came unfiltered from Lake Superior. Dr. Mario Fischer, director of public health in Duluth, testified that "it can be assumed that at no time will Lake Superior simulate a state of complete quiescence comparable to that of a laboratory bottle for a period of thirty days."[30] Fischer pointed out the lake had complex currents, and these currents might carry the tailings straight down toward Duluth.

As evidence, Fischer presented to the commissioners maps of Lake Superior currents that had been created by the early bottle studies. Scientists from the United States Weather Bureau (renamed the National Weather Service in 1970) put a numbered piece of paper into a little bottle, capped the bottle, and tossed it into the lake. Volunteers around the lake gathered up the bottles as they floated by, recorded the numbers off the slips of paper, and sent the data to the bureau. From these early citizen science projects came the first efforts to visualize the currents within the Great Lakes. These citizen science projects of complex circling currents offered a sharp contrast to engineering models of the lake as a big bathtub, a homogenous volume of water so vast that all of Reserve's tailings couldn't pollute it. (As late as 2000, the Army Corps of Engineers referred to the Great Lakes as "like a series of interconnected bathtubs.")[31]

Fischer made larger epistemological claims in the hearings when he objected to Reserve's claims that their tailings would be absolutely safe. We simply cannot know; too much lies unknown beneath the visible surface. He said, "In our opinion, no honest, yes or no, as to whether or not such finer tailings would constitute a nuisance value to the City of Duluth water supply, no honest opinion on that point can really be given." Fischer was troubled because inhaled silica dust—part of the tailings—was known to cause lung diseases, such as silicosis, in exposed miners. In a letter to the City Council, he noted, "It is a well-known fact that the inhalation of silica dust is capable of producing a distinct disease known as silicosis." But because there was no proof that swallowing silica was harmful, he reluctantly admitted that "we find ourselves unable to offer a conclusive or irrefutable reason" for denying Reserve's permits. In other words, the burden of proof had shifted to the state to show harm, not the company to show safety.[32]

In New York a taconite mine had begun dumping its tailings into Lake Champlain, and Wilson wrote to the New York State hygiene board, asking whether the tailings were staying contained where they had been dumped. The New York agency wrote back in March 1947, telling Wilson that the Lake Champlain tailings were causing problems far from the plant: "The amount of settleable material escaping from the separation basin is considerably greater than that contemplated when the application for approval of the plans was made, and an alluvial fan of considerable dimensions has been formed extending out into the lake. Our District Engineer states that turbidity in the effluent has affected the nearby bathing beach."[33] It is possible that Wilson never received the letter, because it appears to have been addressed incorrectly. At any rate, Wilson failed to cite it in his final decision, when he ruled that taconite tailings were unlikely to mobilize into broader spaces.

Fisheries Concerns

Because fishing had become one of the core industries along Lake Superior, tailings' possible effects on fish populations emerged as a key point of contention in the hearings. John Moyle, state fisheries expert, testified that the tailings would likely destroy fish habitat. But if the tailings stayed in the deep trench, as Reserve predicted, fish could just swim elsewhere to escape them. Reserve found some people to suggest at the hearings that tailings might actually improve spawning habitat by offering a new substrate. Reserve suggested that "the deposit of sand does not hurt the fish life [and] some of the places where sand has been deposited in the lake are good fishing grounds in the lake." Moyle disputed this idea, pointing out that trout and whitefish spawned on coarse gravels, not on tailings ground to a fine dust: "These fine tailings will be settling out, and any eggs laid in the coarse tailings would be apt to be smothered in the fine tailings. . . . There is some fish spawning around Beaver Bay, and it is likely that some of the present spawning beds are going to be destroyed."[34]

To make their case that tailings would not harm fisheries, Reserve turned to other mining sites along Lake Superior. In Michigan's Keweenaw Peninsula, copper mines had been dumping their fine tailings, known as stamp sands, into Lake Superior for decades with little oversight or regulation by the state. Engineer Engells from the Keweenaw copper mines testified that at least five copper stamp mills along the shores of Lake Superior had "all deposited their tailings in Lake Superior, and they did so as near as I can find on this particular shore for a distance of about four or five miles, from about 1895, from the first mill, which was the old Atlantic Mining Company." Engells estimated that close to fifty million tons of tailings had been deposited. To him, the lake seemed to him to be doing fine. Anecdotally, he noted that there seemed to be plenty of fishermen out with their nets, which to him suggested that fish were abundant: "Any morning from six to seven o'clock I can see fisherman out there either pulling their nets or setting their nets." In cross-examination, Moyle noted that the Keweenaw copper mines were not an exact parallel because the tailings were a "smaller amount . . . and the stuff was not ground as finely as this will be."[35] Several things are worth noting about this episode: first, much of the Keweenaw Peninsula is now an Area of Concern and Superfund site marked by persistent fish tumors and other contamination; the tailings weren't nearly as benign as Engells suggested. Second, anecdotes became evidence in the hearings when a few casual comparisons became a core part of Wilson's eventual decision to approve the plan.

Conservationists opposed to the Reserve tailings plan suggested very different readings of mining history and its effects on fisheries. They told of a century of development along the St. Louis River flowing into Duluth, largely in service of iron mining, that had destroyed water quality and decimated the once-abundant fisheries. Hundreds of commercial fishermen signed petitions against the project, concerned that what happened in the St. Louis River would happen to their fisheries as well, destroying their economic base. Conservationist Frank Tuskey summarized their testimony: "They realize the fact what is going to happen to Lake Superior and they are down here to do their utmost to protect Lake Superior; and I sincerely believe myself that there should be some precaution taken." The United Northern Sportsmen, a conservation group, brought commercial fishermen to the hearings "who concluded, on the basis of clinkers and debris caught in their nets, that there was a strong current, hundreds of feet deep in Lake Superior, running southwest past Silver Bay toward Duluth." Conservationist Karl McGath described pollution of the St. Louis River, saying that "everybody knows what a rotten deal that is" because the St. Louis River could no longer support fish life. Reserve, he feared, might do the same thing for all of Lake Superior. He asked: "Are we going to take what little bit of power we have—give somebody permission to use the lake for a dumping ground? Does it add up?"[36]

Organized labor came in great numbers to the hearings, and the majority spoke in opposition to Reserve's plan. Thousands of union members signed petitions in opposition if they could not attend the actual hearings. The state dismissed the union members by disparaging them as "communists," according to Thomas Bastow, who served as trial lawyer for the EPA during the Reserve trial in the 1970s. A senior official of the state's Department of Conservation stated, "We looked into who was behind the opposition to Reserve. . . . It was the Communists."[37]

But the evidence hardly supports the contention that opponents of the mine were communists. Union members along the Iron Range were radicalized, but not in support of a communist takeover of the means of production. Rather, they objected to corporate takeover of common property, and they thought Reserve was trying to do exactly that. For example, a representative of one of the local chapters of the United Steelworkers of America came to the hearing and testified: "At our last regular meeting of the Duluth Industrial Union Council held July 15th, 1947, at the Steelworker's Hall, we had a lengthy discussion regarding the dumping of ore tailings into Lake Superior from the new taconite plant from Beaver Bay. The delegates representing all CIO unions in Duluth unanimously went on record opposing the dumping of ore tailings into Lake Superior. We are aware that the dumping of ore tailings into Lake Superior will only establish a precedent for future dumpings of some form of pollution in our beautiful and clear Lake Superior." Similarly, United Steelworkers Local 1028, with 2,711 members, went "on record at the regular meeting held on July 18, 1947, to oppose the dumping of ore tailings or any other form of pollution into Lake Superior at any time." Local 1201 also passed a motion, stating: "We the members of local 1210, U.S.W. of A. are very much opposed to the proposed plans of the mining companies in which they intend to dump the tailings from their taconite plant into the waters of Lake Superior."[38]

While Reserve promised two thousand jobs, the union leadership argued at the hearings that Reserve wanted to dump the tailings into Lake Superior to save on labor costs rather than to create new jobs.[39] Union representatives never spoke against the taconite development projects; they wanted mining, but they didn't want the waste dumped into the common waters of Lake Superior. Because of recent historical experience of logging, they did not trust the corporation's promise to protect common resources, and Lake Superior was the ultimate commons.

The union pointed to the recent past within Wisconsin, Michigan, and Minnesota forests, when large corporations had stolen the northern forest commons, shutting out the public and taking all the profit. The shadow of the cutover loomed large in the mining debates, just as it had in the pulp mill debates. Over and again, people opposed to Reserve spoke of the wreckage fostered by deforestation in the cutover. They argued that only the rich benefited from lumbering, but economic greed devastated the workers and citizens of the cutover lands. Don't let greed destroy our waters the way they destroyed our forests. Conservationist Tuskey testified:

"We must remember this. Years ago the timber barons came, moved in on us. They stripped the land of all the timber, left us nothing but stumps. Where are we now? We are trying to build up reforestation. The mining companies had moved in, which was mighty fine. I like to see industry, but it is a poor industry that can't take care of its waste. It ain't worth while having."[40]

Before effective regulation, Tuskey added, the mining companies had been just as bad as the timber companies, seizing whatever they wanted without regard for the public good. "The mining companies up on the range here had pretty much the run of the land. If they needed timber they went and took it. If they needed water supply they went and took it without asking anybody."[41] This was a common refrain among union members at the hearings: history teaches us that unregulated corporate interests have never protected the common good. The people must stand up for jobs and the environment to protect their future.

A love for Lake Superior was the second reason so many union members and other local citizens passionately objected to Reserve's proposal. Lake Superior had a special value for many people who opposed the mine. They tended to agree that environmental protection might sometimes need to be sacrificed for economic development, but they also agreed that a few special places should remained forever wild. Lake Superior, as the largest, wildest lake in the world, should be one such place. From senators to local fishermen, people from a wide swath of economic and class backgrounds repeated the refrain: Lake Superior is special and worth protecting.

Conservationist McGath testified that the lack of intense industrial development was part of Lake Superior's special value. He said: "Why have we got clean water in Lake Superior? Can you men tell me? Because an operation of this kind has never been on the lake. Otherwise it would be like our other lakes and our other streams where industry has taken over. It would be polluted. . . . We want to keep that clean water just the way we have it in there. It is clean water, and that is the way we want to keep it."[42]

Homer Carr was a member of the railroad union, a resident of Proctor, Minnesota, just outside of Duluth, and a state senator who served from 1935 to 1964. Carr, a long-standing supporter and member of unions, was passionate about economic development for his region. But he was equally passionate about his opposition to the Reserve project. In 1947, he testified, "Now man comes along and says, 'We are going to create a new substance and deposit it in this lake,' and the question is, what price will we pay for violating the laws of nature?" Carr was concerned that this first project would encourage more of the same, until finally the north shore would be "made a gravel basin or disposal basin."[43]

But for some environmental groups, Lake Superior's size meant it could absorb inevitable pollution with less harm than smaller lakes. Lake Superior became a sacrifice zone, in other words, to protect the inland waters. The Izaac Walton League sent two people—Kaupanger and George Laing—to testify that to protect the inland lakes, they would rather Lake Superior be the dump. Laing said of the project: "Everybody in the Isaac [sic] Walton League and most people in Minnesota love the north shore. . . . We would like to support everyone who loves the beauty of Minnesota, fight anyone who hurts. We know we can't here. Our livings come before many other things, and economic values are sometimes more real than our own lives." Kaupanger testified: "We believe that the lesser of two evils would be to utilize Lake Superior as a depository for these tailings rather than some of our beautiful inland lakes." Conservationist Karl McGath, who was also the president of the United Northern Sportsmen, objected to this language of inevitability: "Why give in any lake? . . . It is not absolutely necessary to put it any place so why put it in any lake? So Mr. Kaupanger is willing to trade Lake Superior for an inland lake."[44]

Wilson Makes the Decision

After seven hearings and thousands of pages of testimony, Wilson had to negotiate radically different scientific testimony, contradictory historic narratives, and passionate voices both for and against the project. Science, he had hoped, would provide him with guidance, but he came to realize that scientific understandings were contested. In his summary of evidence at the end of the hearings, Wilson dismissed McGath's fisheries history and instead focused on the tenuous parallels between Keweenaw copper tailings and the presence of fishermen setting nets. Wilson stated: "For over fifty years a somewhat similar, though much smaller, operation has been conducted on the Keweenaw Peninsula in Lake Superior in connection with the Michigan copper mines. The testimony shows that the deposit in Lake Superior of finely ground tailings from this operation for a period of approximately fifty years has had no adverse effect on fishing in this vicinity. The testimony further shows that very similar low-grade iron operations on Lake Champlain in New York State have not damaged fishing in that lake."[45] But Wilson never mentioned the testimony from Lake Champlain showing that their tailings were indeed mobilized into broader spaces, creating turbidity far from the dump site.

The post–World War II era, as described in chapter 3, was when Great Lakes states began concerted efforts to clean up water pollution. As commissioner of conservation in Minnesota, Wilson was in charge of cleaning up the state waters, and he took the hearings seriously. He had led the drive to make Minnesota a leader in water and land conservation. As conservationist Tuskey had pointed out at one hearing, "In the State of Minnesota we have a big program of water pollution . . . if we are going to try and clean up the streams, or one source of pollution, and then turn around and pollute the biggest body of clean water we have in the world, you might say, why, then, I don't see any sense to it at all."[46]

Wilson realized that his efforts to make Minnesota a leader in clean water had to be balanced by the parallel pressures to develop the regional economy. As commissioner of conservation, Wilson had in June 1947, right in the middle of the Reserve hearings, denied Butler Brothers' application to use Oxhide Lake in Itasca County as an iron ore tailings basin. In a letter to the governor written on July 2, 1947—halfway through the hearings and five months before his formal decision—Wilson reassured the governor that even though he had just denied Butler Brothers' tailing permit, he recognized how important taconite development was for the region. He acknowledged that the state had an interest in "promotion of the mining industry on account of employment, tax revenue, and other economic benefits," but the state also had an interest in "conservation of public waters for various public uses. Conflicts between these two types of public interests are inevitable." While Wilson was eager to establish his credibility in pollution control, he clearly felt pressure not to overstep his regulatory authority and stifle economic development on the Iron Range. In this letter, written months before his official decision, Wilson told the governor that he was sure the Reserve project would lead to "practically no loss, and with relatively little impairment of public interests in the lake."[47] These words imply Wilson had made up his mind well before the testimony had concluded.

Three weeks later, on July 29, 1947, Senator Carr wrote Wilson an angry letter stating, "It is my impression that you favor this application." He continued: "I am wondering just how far selfish interests, that control our state, will be permitted to destroy everything of value in our state, under the 'smoke-screen' of more 'jobs.' . . . We know the power of the mining companies, and it is my hope that . . . you will lend your support to preserving the natural beauty of the Lake that is such an asset to our tourist business that today runs to about $200,000,000. Surely this should be considered. . . . The mining industry can deposit this material in some inland swamp just as they do at Coleraine."[48] Wilson immediately responded to Carr insisting that "any impression which you or anyone else may have that this department of the State Water Pollution Control Commission either approves or disapproves any application before a final decision has been made is entirely without foundation. . . . You may be assured that as long as I have anything to say about it no permit will ever be granted that would have the dire consequences described in your letter."[49] Yet his earlier letter to the governor suggests that Wilson had already decided.

After seven hearings, Commissioner Wilson ruled in favor of Reserve, granting a permit for the mine to dispose tailings into Lake Superior. On December 18, the Department of Conservation issued permits; the Water Pollution Control Commission issued permits on December 22, 1947; and in April 1948 the Army Corps issued its permits. Spatial considerations were core to the decisions to issue these permits. Wilson was persuaded by Reserve Mining Company's assurances that tailings would be contained in a very small area within the lake itself. The environmental risks were real, Wilson determined, but he hoped that those harms would remain quite localized—distant from fishing communities and cities along the lake. The economic and security benefits, however, would ripple throughout the state, the nation, and even the globe. Without taconite, the region's economy might collapse; with taconite, a bright future seemed assured.[50]

The permits were subject to three key conditions: first, that the tailings would not discolor the water outside narrowly defined areas; second, that the tailings would not harm fish life in Lake Superior; third, that Reserve would be liable for any harm to water quality. The state reserved the right to revoke the permits if Reserve violated any of its conditions, including an important condition that discharge was not to include "material amounts of wastes other than taconite." Political scientist Robert Bartlett notes that these supposed limits were hedged with weasel words: "In principle, then, the state had imposed severe stipulations on Reserve's use of the lake, but in reality it appeared that Reserve had few worries. The permit only could be revoked for violation of its conditions, but most of the important stipulations were qualified by the weasel word 'material': material quantities, material clouding or discoloration, material adverse effects, material unlawful pollution. Enforcement of the permit would have to be in the face of the reality of an established operation, upon which two entire communities and the welfare of thousands of people would depend."[51]

At one 1947 hearing, a witness opposed to the plant asked Wilson: "How much harder would it be to stop that plant from functioning until such a correction was made than it is now before the application is granted? . . . After that many million dollars have been spent, it is pretty hard to stop an operation." Wilson replied: "It can be done." The critics retorted: "It can't be done. It is pretty impossible."[52] Carr wrote to Wilson with the same concerns, and Wilson assured the senator that Reserve was bearing all the risk. But just as Carr and other critics had predicted, stopping a polluting operation was much harder than preventing one.

Wilson believed that taconite development was inevitable, and like many in his generation, he believed that good conservation meant minimizing the damage, not preventing development. Wilson wrote: "Demands both for water and for tailings basin space . . . are bound to arise in the future for beneficiation of taconite. . . . All this will place a heavy burden on the lakes, streams, and land areas. . . . No one looks with favor on diversion of water from inland lakes and streams or on the construction of inland tailings basins. . . . Yet we know we are going to have to put up with a lot of such inland developments in order to use all the taconite and other low grade ore that is in sight. . . . It is obviously good conservation and common sense to use as much of this vast lake space as can be used and thus lessen the strain on interior resources."[53] For all his caution, in his final record of decision, Wilson essentially ignored most of the concerns raised by opponents. He justified this by defining opponents as non-scientists, people outside the sphere of expert cooperative pragmatism.

Wilson took the state's obligations to protect clean water quite seriously. But he also deeply valued the state's right to make decisions, free of interference, and he wanted to clearly define the state's power and influence. Minnesota's Department of Conservation had only recently, after significant opposition, acquired the right to regulate the mining industries. Wilson had good reason to be cautious about overstepping his still-contested authority. As historian Terence Kehoe argues, the failure of cooperative pragmatism doesn't mean that state agents were in the pockets of industry, nor did the state dismiss the need to protect clean water. But the effort to balance environmental protection with economic growth led to wrenching tradeoffs. Defining Reserve as the true conservationist allowed Wilson to finesse these tradeoffs. Wilson constructed a story about Reserve as an excellent corporate citizen, writing in his decision: "It is true that many corporations, big and little, as well as individuals, are ready to exploit natural resources for their own profit if they can, and we are on guard to stop it as far as the legislature gives authority and the means to do so. However, once in a while someone comes along with a proposition that is right in line with good conservation, and this is such a case. . . . All [Reserve] wants is a chance to prove at its own risk and expense whether or not the proposal to use Lake Superior will work. If it will work without injury to public interests in the lake, it will give a big lift to the solution of the taconite problem, and will relieve considerably the demands which that problem is bound to make on the interior land and water resources of northeastern Minnesota."[54]

In a letter to Senator Carr dated January 12, 1948, soon after the permits were approved, Wilson added that he believed, "The project is in accordance with a sound, long-range program for conservation of both iron ore and water resources, that it will do no material harm to public interests in the lake, and that on the whole it will be in furtherance of the best interests of the state."[55] Wilson knew the growth of the taconite industry might devastate inland water resources. The Lake Superior plan came to seem to him like an almost miraculous end run around a terrible conflict between conservation and economic development. These hopes, and the firm conviction that they could stop the plant at the first sign of trouble, became the basis for his decision. But Wilson soon learned that no matter how much evidence of harm the state gathered, it was much harder to close a polluting plant than it was to keep it from starting up.

FIVE

Mining Pollution Debates, 1950s Through the 1970s

WHEN MINNESOTA COMMISSIONER OF conservation Chester Wilson approved the permits for Reserve's processing facility, reversibility was key to his decision. Wilson acknowledged the limits of knowledge and the flaws with Reserve's assurances of perfect safety, agreeing that no one could possibly know exactly what might happen to the lake or to the health of citizens. Therefore, Wilson promised, the state would grant permits that could be reversed if new evidence of injury emerged. But as Senator Carr and numerous witnesses argued, political pressures and the momentum of development meant that once under way, a billion-dollar project would not suddenly be halted for the sake of a few fish.

After winning permits in 1947, Reserve Mining Company began construction on the taconite processing facility at Silver Bay, on the shores of Lake Superior. The facility's first year of production was 1956, and within a year, new evidence about the mobility and toxicity of Reserve's tailings came to the attention of state regulators. Were Wilson's promises upheld? As evidence of material harm came to light, did Minnesota's state conservation agencies act to protect Lake Superior, or did they follow the lead of cooperative pragmatists in Wisconsin who had tried to negotiate pollution from the pulp and paper industry by calling for more research, more cooperative studies, and more voluntary actions?

In its first five years of operation, Reserve applied for three permit revisions. At each permit revision, the state held hearings, giving itself the opportunity to gather new evidence and close the plant if necessary. The records of these permit revision hearings offer an excellent window into the ways that state conservation agencies tried to incorporate new evidence, both scientific and economic, into pollution regulation.

In 1957, just a year after opening, Reserve applied for a permit to double its water use and its dumping of tailings. Three years later, in 1960, the company asked for permission once again to double tailings deposition and water use. In 1961, the permit was amended a third time to allow the addition of carcinogenic fly ash — a waste product from the electric genera-tion plant — into the tailings.[1] At each of these revision hearings, numerous citizens showed up to describe problems with the tailings. Local environ-mental organizations, commercial fishermen, and sport-fishing groups tes-tified that tailings were killing fish, sliming nets, and clouding the waters.

The first key concern was that the tailings were not staying confined in their deep trench close to the plant. Rather, they were mobilizing far outside the immediate area (figure 5.1). By 1974, tailings had spread over "some 2000 square miles of the west end of the lake."[2] At the time, trying to track the movements of these tailings was extremely difficult because they

Figure 5.1 By 1973, southwesterly currents had carried taconite tailings discharged by Reserve Mining Company's plant in Silver Bay (pictured) to Stony Point, eight miles away, photographed June 1973. (Donald Emmerich, photographer, U.S. Na-tional Archives and Records Administration record 3045077.)

moved underwater. Instead of direct measurements, different groups had to agree on proxy measurements for the movements of tailings that weren't visible on the surface. Not surprisingly, different groups contested these proxies, often bitterly. The lake near the tailings dump had changed in color from blue to green as tailings were added, and the clarity had changed from crystalline to murky. Local fishermen argued that these surface color and visibility changes were good proxies for tailings movements, and the expanding area of green discoloration indicated that tailings were moving into open waters. Yet Reserve contested this interpretation, arguing that nobody could prove beyond a shadow of a doubt that the green murk was caused only by their operations.

In 1957, the hearing record summarized, "Since 1956 — Reserve's first full year of operation — commercial fishermen along the Minnesota shore of Lake Superior had been complaining about stretches miles long, where the clear blue water of the lake was changed to a cloudy green." The Environmental Protection Agency's chief attorney in the litigation, Thomas F. Bastow, noted that the Minnesota Department of Health sent water pollution scientists out to investigate in the fall of 1956 and the summer of 1957. They reported that cloudy "green water" could be seen frequently near the plant, extending from "3 miles northeast of the plant more than 35 miles down the shore to the southwest." The scientists found "much more suspended solid material in samples of this water than there was in samples of clear blue surface water taken from the lake near the Canadian border. When this material was filtered out, it 'strongly resembled the natural gray-green color of taconite.'"[3] For most witnesses, this was conclusive proof that green water was a good proxy for tracking tailings, and taconite tailings were moving far beyond where they were supposed to stay. But Earl Ruble, Reserve's consulting engineer (who had been an employee of the Duluth public health agency during the 1947 hearings), argued that because green water could sometimes be observed away from Reserve, it might simply be natural variation rather than a reliable tracer of tailings.[4]

A delegation of fishermen from Knife River, led by a man named A. N. Ojard, came to the first permit revision hearing to testify about new slimes and silts in their nets that were ruining fishing. Ojard insisted, "There is no doubt about it that some of this stuff does get in the water—these tailings, so to speak. They are very fine and the fishermen have noticed as far as Knife River that even after the nets have been in the Lake one night, they are coated with a white or grayish powder." He noted that the gray powder had never been seen before Reserve opened, but now their mesh "was solid with gray material." Ojard went on to compare the fish with people living in Los Angeles whose lungs were harmed by smog: "I feel sorry for the poor fish. They have the same situation in the water that the poor people in Los Angeles have with the smog condition. If it gets heavy enough, they naturally can't breathe. The fine dust gets in their gills and it has happened out there in Los Angeles. When the smog got thick enough, it gets in there [*sic*] gills; a few did die." He added that fish were struggling with pollution just as urban people were: "The fish, of course, they can't stand the minerals because they breathe through the gills and this fine sediment clogs up the gills and they die."[5] Local fishermen, portrayed by the state and the company as simple fellows with purely local and therefore limited knowledge, were actually paying attention to national news about pollution. They framed their concerns about the health of the fish in terms of human health—terms that everyone could understand, fisherman or not.

A fisherman named Ragnvald Sve agreed with Ojard that tailings were moving far from the plant: "I have seen a considerable lot of this tailing coming up our way all right. I didn't notice too much of it until just here in August [when the plant opened]. That is when I really began to see it . . . the water was so milky colored and greenish . . . I found that in places, you couldn't see more than 2 feet down. Ordinarily, Lake Superior, when the water is clear and normal, you can see about 30 feet down."[6]

Rather than ignore the testimony of fishermen, the state paid attention to their concerns, ordering Reserve to send scientists out to investigate these slime concerns. Reserve engineer Ruble reported back with exactly the same response he gave to reports of green cloudy water: the gray slimes in fishermen's nets had "no connection whatever with the operation of the taconite plant at Silver Bay. . . . It's a perfectly natural occurrence."[7] Rather than accepting this without question, the state sent its own scientists to investigate the slimes, and they found ample scientific support of the fishermen's claims. Far more gray slime was found on nets they set for a single day near the plant than on nets they set for ten days near the Canadian border. They analyzed the slimes with microscopes and X-ray diffraction, reporting that taconite tailings were in them, encouraging the growth of algae and iron-fixing bacteria. There was little doubt that the tailings were moving up and down the lake shore and into the open lake—and little doubt that they were causing slime problems in fishermen's nets.

As part of these investigations, in 1957 the state's consultant scientist (John Gruner from the University of Minnesota) identified asbestiform minerals in the tailings from Reserve. He also found those same asbestiform fibers in the discolored waters of Lake Superior where tailings were mobilizing. Gruner identified a major constituent of the solid material in the "green water" samples and in the tailings themselves as a mineral of the amphibole group called cummingtonite-grunerite. He also reported that this asbestiform fiber was due to Reserve's operation, because it "was absent in the control samples taken near the Canadian border. It was also absent in a sample of lake sediment taken at Silver Bay before Reserve began operations." Professor Gruner concluded that his samples certainly came from the eastern Mesabi Range, where Reserve's mine was located (see figure 4.2). These should have been revolutionary results, but the state suppressed the results for more than a decade, according to Bastow. Even with multiple lines of accumulating evidence, the Minnesota Department of Conservation continued to tell the public "that there was no evidence of pollution from Reserve's discharge." Bastow notes that one state water pollution scientist later explained, "It was the philosophy of the Department at that time not to push."[8]

In 1960, during the hearing regarding the second permit amendment, the state admitted the possibility of asbestos in the tailings, writing that "an identifiable mineral constituent of taconite known as amphibole" had been found "in some of the slime material taken from the fishermen's nets in the Silver Bay area." But rather than clarifying that these amphibole fibers were potentially asbestos, the state diluted the power of this statement. The state refused to use its powers to intervene, even though permit conditions had been violated.[9]

Wilson had promised in 1947 that Reserve would bear all the risks and would assume the burden of showing its operations were safe. But by 1957, the Minnesota Pollution Control Agency came to interpret the initial permits as meaning that petitioners against Reserve bore the burden of proof. It wasn't enough for locals to show that pollution was resulting and that it was moving past the confined spaces allowed in the permit. Finding toxic chemicals in samples wasn't sufficient. Evidence that fisheries were collapsing also wasn't enough. Instead, the state came to require that petitioners show proof that pollution was from the mine and causing the fishery problems. Proving such causal links, however, was nearly impossible.

Science and technology studies scholar Barbara Allen's research explores the challenges in changing older technologies when problems emerge *after* they have become entrenched.[10] A lack of research was never the problem for governing Reserve. The state had organized careful monitoring regimens as soon as the plant opened, and when the state received complaints from the public, state scientists investigated rather than ignoring concerns. But when new information emerged from state research, regulators were unable to correct course and integrate that new information into policy. Using monitoring results to change policy was almost impossible, given political pressures on the state to encourage industry.

A huge amount of scientific data gathering was required under the terms of the initial permit. Yet the key question for better environmental policy making should not be: Is monitoring required? Rather, it should be: What happens with that monitoring information? Under what conditions do large institutions—whether state regulatory authorities, corporations, or federal agencies—have to shift course? The state had three opportunities in the first five years to collate a rich diversity of monitoring data, interpret them, and then modify behaviors in response to that new information. But for such adaptive management to work, several conditions must be in place: First, transparency of information must exist. Whoever gathers the data must be willing to share all the data and not just cherry-pick the data that support their case. But the hearing records and internal Reserve memorandums show that the company shared only the data that supported their case, suppressing other data.

By the late 1940s and early 1950s, scientists and medical researchers required controls, blinded analyses, and careful sampling to address cognitive biases that lead many people to ignore data that don't support their preconceived ideas. But regulatory agencies in Minnesota didn't seem to be aware of the importance of monitoring the monitors. Rather, they increasingly trusted Reserve to do the right thing. When Wilson positioned Reserve as a good conservation citizen, he created an intellectual and moral space that assumed the corporation had the lake's best interests in mind. Fearing that challenging Reserve might undermine their cooperative pragmatist models of pollution control, the state was careful not to question Reserve too closely.

One example of this is the debate over fly ash, a toxic residue produced at the taconite plant. Reserve's energy-demanding beneficiation plant required a great deal of coal fed into the electric power plant, and coal burning produced fly ash. The original permit explicitly prohibited fly ash from being discharged into the lake. But on December 1954, nine months before the plant was to open, consultant engineer Davis wrote to the vice president of Reserve warning him that fly ash was being combined with the tailings and would get into the lake. Davis then suggested that Reserve should not report this problem or attempt to fix it unless "trouble develops."[11] Even before the plant opened, problems emerged that the company suppressed. Eventually, Reserve asked the state to allow the company to discharge toxic fly ash from its electric power plant into the main tailings launder, and from there into Lake Superior. The Conservation Commission granted Reserve permission on January 30, 1961, assuming that fly ash, while extremely toxic, would be so dilute that it would not hurt aquatic organisms.[12]

By 1966, Reserve's staff were suppressing information about fly ash effects on fish. On February 24, 1966, Edward Schmid, director of public relations and assistant to the president for Reserve, wrote an internal memo regarding the application for a renewed fly ash permit. Reserve had submitted a small sample of information to the state purporting to show that fly ash was not visible or measurable in lake water, carefully selected from certain months where such ash would not likely be visible. Schmid warned

Reserve staff not to mention that Reserve had not measured during other months: "I suggest we remove the statement that admits Reserve does not have similar readings for other months of the year. It is not necessary here and it is a written admission we prefer not to make."[13] The next year, Reserve corresponded with engineers from a power company to determine the effects of fly ash on fish, and they received a full list of scientific papers discussing its toxic effects. In internal memos, Reserve noted that their staff had read these papers and discussed evidence of harm to fisheries, including one "recent accidental case of a fish kill in which fly ash was involved. The fish kill occurred when a dam on a fly ash holding pond broke." In other tests, all test fish quickly died, and "death seems to come from oxygen lack, i.e., asphyxiation, because the gill mucus becomes clogged by ash particles."[14] Rather than provide this information to the state, the company continued to deny any knowledge of fly ash toxicity.

Handwritten field notes in the Reserve files show that Reserve knew its fly ash samples contained toxic materials at significantly higher levels than allowed by their permit. But rather than report these findings to the state as required, Reserve continued to tell the state that its fly ash presented no problems. None of this is surprising; other studies have found similar patterns of behavior when corporations suppress knowledge about environmental harm while insisting to regulators and the public that their processes are completely safe. What's surprising, perhaps, is that the conservation agencies took so long to realize that their partners were not completely honest.[15]

Taconite Booms

By the late 1950s, the taconite boom that Reserve stimulated was having a profound effect on the region's economy, just as the pulp and paper boom had stimulated the Canadian economy along Lake Superior (figure 5.2). Towns thrived; new cars filled the parking lots where well-paid union workers toiled in the taconite facilities. Schools improved, funded by abundant tax receipts. On the Keweenaw Peninsula, copper mining's collapse had left a devastated economy. In the Gogebic and Penokee Ranges, deep-shaft

Figure 5.2 Hull-Rust-Mahoning open-pit iron mine in 2014, from mine overlook, Hibbing, Minnesota. This mine stimulated an economic boom in the region. (McChiever, via Wikimedia Commons.)

iron mines had shuttered with competition from the taconite mines in Minnesota. But on the iron ranges of Minnesota, taconite delayed for several decades the economic decay that plagued other mining areas around Lake Superior.

Investment in taconite mining and processing soared. Other companies including U.S. Steel constructed mines on the Mesabi Range, building taconite beneficiation plants close to their mines, even as Reserve continued to send its taconite by rail to the Silver Bay beneficiation plant for processing. One trade paper wrote in 1967: "The huge but worn-out Mesabi iron range is being revitalized by continuing construction of facilities for processing its deposits of magnetic taconite. Early next year U.S. Steel's 4.5 million ton-per-year taconite pellet plant at Mountain Iron, Minn., will reach full production. . . . To date, more than $1.3 billion of capital has been spent on plants to produce iron from taconite ore." Taconite became a symbol in Minnesota, for progressives and conservatives alike, of the prosperity that technology could bring. Hubert Humphrey, author of the Civil Rights Act, vice president under President Johnson, and Democratic nominee for president, was perhaps Minnesota's most famous progressive politician. In 1968 one paper reported, "Hubert Humphrey last week compared the untapped resources of our nation's poor people to taconite being extracted from Minnesota's worked-over Mesabi iron range." When Humphrey said, "There are hundreds of thousands of people in the form of human taconite waiting to be utilized and developed," he captured some of the optimism for a more prosperous future that taconite had come to symbolize.[16]

Most basin communities around Lake Superior still believed the water was so clean that there was no need to filter it before drinking. But as communities grew, concern about the effects of Reserve's taconite tailings on the lake grew as well. Political scientist Robert Bartlett calculates that in 1968 184,000 people each day got their drinking water from Lake Superior, withdrawing a total of twenty-five million gallons per day for municipal use. That amount was dwarfed by the half billion gallons per day that Reserve was withdrawing from the lake near Silver Bay. One beneficiation plant was using twenty times the water consumed by the human population around the lake.[17] But harms to the lake from the taconite boom were subtle and hard to pin down whereas the benefits were clear to see: jobs for miners and economic development for the region.

Throughout the 1950s, as evidence accumulated that taconite was causing pollution problems, state agencies continued to insist that the industry was harmless. But federal researchers were beginning to be concerned. Commissioner Wilson had long been concerned about the balance between state and federal power. His 1947 decision had come just months before passage of the Federal Water Pollution Control Act of 1948, which provided a limited federal role in clean water regulation. Because water pollution was still considered a state and local problem, the 1948 act provided some assistance to local government, but no federal regulatory power — or even water quality guidelines.[18] Even with these limits, Minnesota was fiercely opposed to any federal involvement in water quality governance.

Senator Gaylord Nelson from neighboring Wisconsin became increasingly frustrated by Minnesota's reluctance to challenge a key industry.

Nelson was particularly concerned that the pollution from Reserve might be crossing into Wisconsin's Apostle Islands National Lakeshore, twenty-four miles across the lake. In 1963 Senator Nelson challenged Murray Stein, the federal water quality act enforcement chief, to take action against Reserve, asking him: "What about the pollution in Lake Superior caused by mining, pouring pollution into Silver Bay from the big taconite mines in Minnesota?" Stein replied that the pollution remained within a single state, so it wasn't within federal jurisdiction. Nelson disagreed, pointing out that the lake itself constituted interstate waters. Stein insisted that the pollution had stayed in local spaces close to the plant, completely within Minnesota waters, using as evidence a brief flight from a spotter aircraft that seemed to show green water only close to the plant. Stein told Nelson: "We have looked at that taconite discharge and I think you can see it from an airplane. Generally it stays relatively close to shore and drops out." In 1971 David Zwick and Marcy Benstock, two members of Ralph Nader's study group on water pollution, commented: "For more than five years after Stein's sub-committee appearance the Federal government considered that view from a spotter aircraft sufficient proof that none of Reserve's taconite tailings were wandering into Wisconsin waters."[19]

The next federal expression of concern came from a federal water pollution laboratory in Duluth, established with the help of U.S. Representative John Blatnik of northeast Minnesota. Blatnik's complex ties with Reserve illustrated the dilemmas that taconite pollution created for conservationists. Blatnik was a towering liberal in Congress, famous for supporting unions and economic development that benefited the Iron Range.[20] He had also long been known as "Mr. Water Pollution Control," for he promoted the federal pollution control program and was author of all the water pollution legislation enacted by the House during his time in Congress. Blatnik was good friends with Reserve company officials, particularly president Edward Furness, and he had long been known as "a booster of the taconite industry—backbone of Northeastern Minnesota's economy and his district's largest employer."[21] By the 1960s those two roles came into conflict.

With Blatnik's help, in 1967 Duluth got a national laboratory devoted to research on water pollution (the National Water Quality Laboratory).

Blatnik intended the new National Water Quality Laboratory to be a show-piece of the federal government's interest in water pollution control. One of the first scientists transferred to the new lab in Duluth was an aquatic biologist named Louis Williams who had decades of experience studying eutrophication. When he arrived in Duluth, Williams visited with local fishermen, who told him about the pollution from Reserve's discharge into the lake. His first assignment was to survey the limnological properties of Lake Superior water, and "he found almost immediately that the water was not as pure as everyone had believed." He began driving along the north shore, taking water samples from the lake and its tributary streams. And then, back in the laboratory, he added different concentrations of tailings to his lake water samples, to quantify the effects of tailings on algae popula-tions. William's experiments showed that Reserve's tailings were "stimulat-ing blue-green algal growths, particularly at higher water temperatures. . . . And in the Duluth water supply Williams had discovered algal conditions that suggested a gradual trend toward Lake Erie conditions."[22]

When Williams told his supervisors at the lab, they were deeply con-cerned—but about what might happen if Williams's finding were publi-cized, not about the pollution itself. After Williams gave a speech to a local civic organization about Reserve's discharges, he was warned to stay away from "sensitive" research areas and stick to basic limnology. His speech had come to the attention of Leon Weinberger, an assistant commissioner in the Federal Water Pollution Control Administration, who complained that Williams "had no clearance" to speak about Reserve.[23] (Weinberger, it's worth noting, was yet another revolving door regulator who soon left federal employment to consult for Reserve.)

Williams found his research stymied at every turn. The lab's acting director told him that he couldn't publish anything without clearance, yet for mysterious reasons, he could never get clearance—but he also got no firm reasons for denial of publication. Frustrated, on June 28, 1967, Wil-liams wrote to Senator Nelson: "Since I am already in trouble for letter-writing—one more might be a public service and probably cannot hurt me more. . . . The question that I would like to raise concerns the pollution of Lake Superior (eutrophication) with taconite wastes from Silver Bay,

Minnesota. . . . We must not allow Lake Superior to become another Lake
Erie." Williams told Nelson that "his work had substantiated charges that
Reserve Mining Company was polluting Lake Superior." Williams also
wrote to the Wisconsin conservationist Martin Hanson, who released Williams's letter to the press in September 1967. The letter did not mention
Reserve Mining Company by name, but it noted that "preliminary studies
seem to indicate that we have a basis now to be gravely concerned about
the deterioration of the water quality of Lake Superior." The letter claimed
that Reserve's taconite could be found across state boundaries in Wisconsin's Apostle Islands, making the pollution a federal matter.[24]

    Williams's letter sparked fury. Across the border, Wisconsin Gover-
nor Warren P. Knowles objected to Minnesota's pollution harming Wisconsin waters, and the attorney general of Wisconsin denounced Reserve
in the media. The head of the Wisconsin Department of Natural Resources
wrote to the new director of the Duluth lab, Don Mount, asking him how
he was going to follow up on Williams's research. Mount replied by under-
mining Williams's credibility, writing, "I feel this is an irresponsible report
based, insofar as we know, on 'bankside' observations, and is embellished
with generous speculations.[25]

    Mount then wrote to Williams, warning him that he was playing with
fire by releasing preliminary results: "I think you will agree with me that fire
is dangerous for young children to play with. The same is true in regard to
the non-scientific people and their ability to understand 'preliminary studies.' . . . Show me the actual data that you have collected which are strong
evidence that there is a cause and effect relationship between deterioration
of Lake Superior and pollution sources. Further, furnish the data which
are evidence that the deterioration now . . . is due to pollution and not
to natural aging of the lake."[26] Note that Mount didn't dispute Williams'
conclusion that the tailings were polluting the lake. Instead, Mount urged
silence about these results until they could be replicated and until the lab
could carry out additional experiments proving that taconite tailings were
the cause of algal blooms, eutrophication, and fisheries decline. Williams
refused to be silent, pointing out that the burden of proof should not be on
the regulatory agencies.

Williams quit his job at the federal lab and published an article about the taconite pollution in *BioScience,* one of the nation's leading scientific journals. The article contained preliminary data about trace metal contamination from taconite tailings, using those data to argue that if trace metals were the limiting factor in Lake Superior algal growth, tailings might lead to terrible algal blooms throughout the lake, turning it into another Lake Erie. Williams added, "These preliminary studies seem to indicate that we now should be gravely concerned about the deterioration of water quality of the Great Lakes from trace metals from colloids, and their effect on producing blooms of nuisance blue-green algae." The major contribution of the *Bioscience* paper was not in its limited scientific data, but rather in its forceful comments about the importance of preventing another Lake Erie crisis. Williams wrote: "We are aware that the conversion of low grade taconite rock into usable iron ore is an expensive operation and must be competitive. The removal of the added trace metals, however, from Lake water is even more expensive." Rather than let Lake Superior, still relatively clean, follow the same pattern of collapse as the other Great Lakes, Williams urged policymakers to pay attention to early warnings of harm. He wrote: "Other writers are now saying that Lake Erie is dead and that Lake Michigan is sick. Along these lines I would state that Lake Superior is now infected."[27]

Regional and national media picked up the story. One man named Ernest Swift wrote a piece in *Conservation News,* a publication of the National Wildlife Federation, stating that one of Reserve's engineers had told him, "Sure we kill a few fish, so what." Swift urged lawmakers to act, writing: "Lake Superior need not be despoiled with filth. . . . There are alternatives in handling these wastes by ponding them, but this will cost money which industry will not spend unless forced to. It would, however, save many a midland waterway and lake from utter destruction."[28]

Reserve was furious, and the head of its public relations team, Edward Schmid, sent an internal memo to Reserve president Furness outlining their strategy for silencing the critics. He wrote that he had persuaded a writer for the *Minneapolis Star* to suppress an editorial about the pollution allegations: "After a very long explanation and discussion, I was able to convince him the article was not factual and he told me he would not write an editorial based on it as he had planned. . . . I telephoned other editors and friends in the area, explained the situation, and got agreement from editors not to publish the material—without checking with us, at least."[29] Schmid next detailed the political favors he planned to call in to quell the story. Concerned that the National Wildlife Federation might speak out against Reserve, Schmid decided to investigate the executive director, one Thomas Kimball, for possible wrongdoing that could be used to silence him in the press. He contacted Bob Tuveson, chairman of the Minnesota Pollution Control Agency, to complain about media coverage. Then he arranged for Reserve's attorney to contact Congressman Blatnik and get him to work on Kimball.

Reserve's reaction to a few letters appearing in regional papers might seem extreme, but the intensity of its response reflects a growing concern within the taconite industry that federal pressure might come to bear against them. Schmid noted in an internal memo to Furness that "the situation is touchy" because of an upcoming permit review from the Army Corps of Engineers.[30] To understand why this worried Reserve so much, we need to step back for a moment and consider changing federal control over water pollution.

As Terence Kehoe argues in *Cleaning Up the Great Lakes,* before the late 1960s, states and municipalities had made little progress regulating polluters because they tended to form cooperative boards and commissions that sought common ground. Companies typically responded to the possibility of new regulations by threatening to pack up and find a more congenial place to operate.[31] These cozy relations between state regulators and corporations began to be challenged in 1956 when the Federal Water Pollution Control Act was revised to allow for enforcement conferences to address pollution in interstate waters. Under provisions of the 1956 act, the federal Public Health Service and representatives from the states affected by interstate pollution would consider evidence at a formal conference. If the conference found that pollution was crossing state lines, polluters would be notified that they had six months to address the problem. If the polluter failed to act, another public hearing and another six-month notice would result. Only after the second deadline expired would the federal government take the polluter to court—and only if the state where the pollution originated consented. It was an awkward and unwieldy process, but it had the potential to call national attention to pollution problems.

Federal leverage to address water pollution increased again in 1965, when the Federal Water Pollution Control Act was amended to require each state to adopt water quality standards in compliance with federal minimum standards.[32] The act authorized the newly created Federal Water Pollution Control Administration to set pollution standards when states failed to do so. This law allowed the federal government to intervene in Reserve's pollution release, but only if the parties could prove that Reserve's pollution had left state waters. In late 1967, both the U.S. Army Corps of Engineers and the Department of the Interior agreed to review Reserve's permits to make sure they were in compliance with the new federal water quality standards. Minnesota, however, vigorously contested federal intervention into Reserve's pollution.

The Army Corps played a significant role in the Reserve case. In 1967, a memorandum of understanding had been approved by the secretary of the army and the secretary of the Interior that established cooperation between the two federal agencies. This enabled biological concerns, not just navigation concerns, to be part of federal permit renewals by the Army Corps. On November 1, 1967, the Army Corps issued a public notice that it would revalidate Reserve's permits unless objections were filed by January 31. Minnesota agencies wrote back that they supported revalidation, with the Minnesota Department of Conservation stating that it had no evidence tailings were seriously harming fish. Wisconsin's Department of Natural Resources, however, immediately objected to Reserve's permit revalidation on the grounds that tailings might be harming fisheries in Wisconsin's Apostle Islands.[33]

The Stoddard Report

The request by the Army Corps and Wisconsin's response persuaded Stewart Udall, secretary of the Interior, to create a task force to investigate

Reserve's pollution. Charles Stoddard, Interior's regional coordinator, was chosen to serve as head of the task force, with members coming from five different agencies within the Department of the Interior.

Born in 1912, Charles H. Stoddard grew up in Milwaukee. He became a conservationist quite young, attending Izaak Walton League meetings with his father. He trained as a forester, and after serving in the South Pacific in the Naval Reserve during World War II, he joined the U.S. Forest Service as an economist for the Lakes Region, arguing for less clear-cutting on federal lands. He next became Bureau of Land Management director, where he became embroiled in controversy when he increased grazing fees charged to ranchers. According to one federal scientist who worked with Stoddard, he was "a real 'blue sky-er'—he believes in zero discharge as an industrial standard." In 1967, Stoddard "ran afoul of powerful timber interests who then essentially controlled the Forest Service." Tired of Washington, Stoddard returned to the Midwest, close to his family's cabin in northern Wisconsin.[34] Stoddard was friends with two Wisconsinites who were powerful voices for Lake Superior: Martin Hanson and Gaylord Nelson, who told him of Louis Williams's experience with Reserve.

Because Stoddard had worked with Udall and other Washington politicians, some Lake Superior conservationists feared that Stoddard's report would be a whitewash vindicating Reserve. But Stoddard prided himself on his scientific meticulousness, and Nelson's and Hanson's suggestions that Reserve had silenced Williams's scientific findings made him intent not to let the same powers silence his research. In early 1968, Stoddard held numerous meetings with staff from multiple agencies to collate data about Reserve. He included Reserve employees and staff from Minnesota agencies in the meetings, although the Minnesota Pollution Control Agency staff were reluctant to participate. Stoddard recalled that he began to suspect Reserve was hiding evidence of pollution when the lawyer for Reserve, Bill Montague, "hounded us all the way through the committee's work." Stoddard said that Reserve's presentations to the task force "just didn't square with the evidence we were gathering. It was bullshit." For example, Montague's formal statement to the task force insisted that fisheries had not been affected in the slightest, even though several formal studies showing significant effects on fish had already been presented to the committee. Instead, Montague insisted, any observed declines were due entirely to the sea lamprey (which were a significant concern), not to pollution.[35]

A key study considered by Stoddard concerned tailings' effects on aquatic food chains that supported lake trout and whitefish. The study showed that taconite tailings suppressed populations of a genus of zooplankton then known as *Pontoporeia* (now *Diporeia*), an important base of fisheries food chains. The paper estimated that Reserve's tailings would result in an estimated loss of 5 percent of the lake's total fish production per year. Reserve took this estimate, assumed it applied only to the nine square miles directly in front of the plant, and trumpeted that the fish effects were completely trivial. The state of Minnesota repeated these narrow claims in objecting to Stoddard's report, insisting with Reserve that "the economic loss in fish production attributable to the present deposition of tailings . . . is quite minor—less than 5 percent of the annual harvest of fish. . . . These values should be taken into account when considering any possible suggested alterations in plant operations."[36] When this estimate of a 5 percent decline was applied to the entire lake, it meant a substantial impact on the fisheries of the world's largest lake. Yet the state insisted on interpreting this evidence in very local terms, assuming that the pollution would stop immediately in front of the plant, even though a decade of research had shown the opposite.

A draft of Stoddard's report was sent out for peer review by state and federal scientists, with the final version incorporating their comments. John Moyle, Minnesota's lead state fisheries biologist, praised the body of the report but expressed concern about the conclusions, worried that the uncertainties that surrounded the data sets were not qualified carefully enough in the conclusions. The review process meant that the Minnesota agencies were not taken by surprise by the final report, as agency leaders later claimed. For example, the draft included a surprisingly high arsenic reading reported to Stoddard's team by the Federal Water Pollution Control Administration. The head of Minnesota's conservation agency, John Badalich, wrote to Stoddard objecting to this data point, and Stoddard immediately agreed to have an independent company reanalyze the samples. After the lab found an error in the original arsenic results, Stoddard dropped them from the final report. In other words, the review process worked.[37] But this temporary error—not surprising in a draft that collated data from multiple agencies—became the basis for a continued effort to slander Stoddard and discredit the report.

Stoddard sent the revised draft report to Don Mount, lead scientist at the federal lab in Duluth (who had earlier scolded Williams for his research on taconite pollution). Two years earlier, Mount had essentially agreed with Reserve that Lake Superior was so big, so clean, so low in nutrients, and so cold that a little pollution couldn't do much harm. But in October 1968, after reviewing the copious data that Stoddard had gathered, Mount reversed himself, arguing that Lake Superior was uniquely vulnerable to pollution. Mount wrote a series of conclusions for the Stoddard Report recommending that Reserve "be permitted to continue depositing tailings in Lake Superior only on the condition that Reserve construct on-land tailings disposal facilities and recycle its waste water, within three years."[38]

After the review and revisions, Stoddard provided Reserve, as well as the five Interior agencies, with a copy of the report on December 31, 1968. Thomas Bastow, the EPA lawyer, noted that "the report's conclusions were shocking." Tailings contained a ton of nickel, 2 tons of copper, a ton of zinc, 3 tons of toxic lead and chromium, 25 tons of phosphorus (a key nutrient leading to eutrophication), and 310 tons of manganese—for just a single day's deposit into the lake (figure 5.3). Silica, arsenic, and of course iron were also part of the waste dumped into the lake each day. The lead, iron, and copper concentrations violated federal law. The company had promised the tailings wouldn't move from their deposition site, but Stoddard found that only the coarsest materials settled to the bottom—perhaps 45 percent of the total. The rest were ground so finely that they remained suspended in the water, traveling with currents around the western arm of the lake and contributing to the lake's eutrophication. More than a third of the tailings had simply vanished from the company's accounting, mysteriously transported by currents throughout the western arm of the lake (and perhaps farther).[39]

Figure 5.3 Water flowing across the fine taconite tailings mobilized heavy metals, asbestos, and sediments into Lake Superior from Reserve Mining Company's processing facility. The EPA caption to this photograph from June 1973 reads, "Foul smelling steam rises from the mucky water. Reserve's taconite plant at Silver Bay produces iron ore from the taconite rock. It discharges 67,000 tons of waste rock into the lake daily." (Donald Emmerich, photographer, U.S. National Archives and Records Administration record 3045077.)

The report's recommendations were grounded in a nuanced reading of limnological history. The problem with pollution was not just that it might add toxics. Rather, as Stoddard noted, the tailings were "possibly accelerating the rate of aging of the lake." Lakes change, but toxics altered these processes of change: "Once metals such as copper and zinc are put into the lake, and because the turnover time of the lake is exceedingly slow, they are going to be there for a long time, and they will accumulate." Stoddard noted that Reserve's pollution did not occur in isolation; the watershed as a whole needed to be considered, and cumulative effects of many small stressors might tip the lake into a different ecological state. He wrote: "Changing conditions may occur in the lake from the addition of other pollutants by municipalities, paper mills, or from materials carried into the lake by the tributary streams. Should conditions later change, it will then be too late to remove the pollutants from Lake Superior and a very serious problem could exist."[40] Stoddard did not portray the lake as a static, unchanging entity to be protected in its current state. Rather, the lake was part of broader ecosystems that existed in a historical context combining natural and anthropogenic processes.

After Reserve received the Stoddard Report, its staff immediately placed pressure on Secretary Udall, urging him not to release it to the public, arguing that it was "riddled with errors." The company hired a series of consultants to attack Stoddard's credibility as a scientist and a person. For example, one consultant scientist, Fred Lee from University of Wisconsin, wrote to the Federal Water Pollution Control Administration charging that the Stoddard Report "was a highly biased attempt by Mr. Stoddard and those that assisted him in writing it to excite the public into forcing the Reserve Mining Company to change its method of disposal of taconite tailings to Lake Superior. I shall always attempt to be on the other side of any group who try to use tactics such as those in the 'Stoddard Report' in an attempt to prove water pollution . . . it is technically unsound and an irresponsible piece of reporting by Mr. Stoddard. . . . I will not stand by and let the public be further misled by the 'Stoddard Report.'"[41]

In a prepared statement released to the media, Schmid asserted that the Stoddard Report contained serious errors, misinterpretations, and unsupportable conclusions. He referred to the report as a "working draft," an "erroneous paper," and a "purported summary report." Federal and state agency staff echoed company language in their attacks on Stoddard. John Badalich of the Minnesota Pollution Control Agency echoed the words of Reserve's press release, claiming that the report was rigged from the beginning, with numbers "manipulated to demonstrate a preconceived notion of the situation" and conclusions based on "speculation and conjecture designed to arouse public concern."[42]

Schmid called Congressman Blatnik's office "to express his outrage at the findings." After that phone call, Blatnik began to undermine the Stoddard Report, using Reserve's talking points and calling it "a preliminary report with no official status . . . riddled with mistakes." Next, Max Edwards, Udall's assistant secretary of the Interior for water quality, called Stoddard and ordered him to Washington, telling him that Blatnik was "disturbed" about his report. Edwards attacked the report in the media, again using Reserve's words, calling it "unofficial" and filled with unspecified "errors in conclusion." Edwards placed increasing pressure on staff to suppress the report (until he quit the federal agency and went to work for a grateful Reserve Mining Company). Even Udall himself initially refused to support Stoddard, telling Congress that the study was "a preliminary staff report."[43]

Stoddard fought back, writing letters to Gaylord Nelson asking for help. Afraid that the federal government would suppress his findings, Stoddard spoke directly to the media, and soon a *Washington Post* editorial charged that Blatnik had suppressed the report. Editorials attacked Blatnik's reputation as Mr. Water Pollution Control. Even with the media attention, the Federal Water Pollution Control Administration attempted to bury the Stoddard Report. The agency ordered a revised document to be prepared and presented as evidence at the federal enforcement conference in May 1969; the original version was supposed to be reduced to the status of a draft. The revision followed Reserve's talking points by removing numerous details about how far the tailings had mobilized into the lake; about the connections between discolored water and tailings; about declines in zooplankton near the tailings; and about toxics near the plant that violated federal water quality standards, particularly lead, copper, and chromium. Reserve internal memos reveal how pleased the company was with the official version, particularly because it removed the recommendation that tailings disposal into Lake Superior be ended within three years. Instead, the revision called for "surveillance" and "further study"—two results Reserve could live with.[44]

At the federal enforcement conference, meant to determine whether Reserve's tailings were polluting interstate waters, the censored revision, but not Stoddard's actual report, was included in the formal record. Conservationists were furious. Gaylord Nelson was too ill to attend the conference, but his prepared statement was read into the record. He focused a great deal of attention on the possible human health effects from Reserve's pollution, arguing that copper, lead, and cadmium levels near the plant were "all beyond the levels of safety for drinking water." After noting reports from commercial fishermen that "the decline in lake trout and herring has not been reversed by sea lamphrey [sic] control," he argued that taconite tailings might contribute to lake trout collapse. The speech contained a repeated theme of the need to learn from history: "Shall we let Lake Superior in a few short years become as degraded as Erie and Michigan? Or do we use our God-given knowledge gained from these past tragic mistakes in Erie and Michigan by taking intelligent action now to save this largest body of clean, fresh water in North America?" Cleaning up pollution costs money, Nelson admitted. But "it is cheaper to clean up the sources of pollution now than to postpone another 5 or 10 years and have the problems of Lake Michigan and Lake Erie facing us."[45]

Nelson kept the pressure on the federal government to regulate Reserve, writing to Walter Hickel, the incoming secretary of the Interior, that Reserve's pollution "can spell doomsday for the lake." He reiterated his frustration about the failure to move against Reserve, writing: "If we fail to deal with this polluter, it will cast a shadow over the entire national pollution control effort. Every industry and municipality, large and small, will have the right to say: 'You didn't make them stop. You can't make me.'"[46] Allowing Reserve to pollute Lake Superior would set a terrible precedent, Nelson believed.

After the enforcement conference, Reserve's position began to crumble. Yet the company kept working hard to keep the edifice together, using calls for more scientific research as their major delaying tactic. Conservationists were furious, with one person writing to the *Minneapolis Star,* "By the time the politicians and mining company lawyers get around to doing anything about it, the sea of red tape will be larger than Lake Superior, and we will be able to drive across taconite tailings from Silver Bay to Ashland, Wis."[47]

Reserve's manipulation of scientific expertise and evidence became clear throughout the next year, when the state of Minnesota finally turned against it in court.[48] Recall that in 1965, the Federal Water Pollution Control Act had been amended to require that each state adopt water quality standards, to be approved by the federal government. In 1969, Minnesota completed this process, and the federal government approved their new standards. A month later, Reserve took Minnesota to court over the approved standards, arguing that the state had no right to apply the water quality standards to Reserve's operation.

In retrospect, the 1970 trial was a fatal move for Reserve because it meant that the company finally lost the support of the state agencies. Furious that after years of state support, Reserve had sued the state, Minnesota attorney general Douglas Head counterattacked, asserting in court that Reserve was indeed "polluting Lake Superior in violation of Minnesota pollution control statutes and regulation."[49] Head went so far as to claim that Minnesota had been "long committed to stopping the dumping of these taconite tailings into Lake Superior"—a bizarre statement, given the state's efforts to malign Stoddard and stop the federal enforcement conference. At trial, for the first time, Minnesota framed Reserve as a terrible polluter rather than an ideal corporate conservationist.

Reserve found a long parade of scientists to testify against the state, focusing on the questions of tailing transport. Reserve insisted that it was impossible to trace the movement of tailings and therefore impossible to blame the pollution on Reserve rather than on nature. These were familiar arguments, but the testimony provided the court evidence that Reserve consultants were knowingly and intentionally manipulating data.

Dr. Alfred Beeton, a limnologist from the University of Michigan, testified that Reserve was intentionally offering poor data about water chemistry and algal growth to confuse the court: "Because of invalid assumptions about Lake Superior, Reserve's water chemistry and algal growth studies contain no data from which conclusions about Reserve's discharge can be drawn. Algae populations have increased significantly in the western arm of Lake Superior since Reserve commenced operations. Reserve's discharge is contributing materially to the accelerating eutrophication in Lake Superior." Beeton accused Reserve of choosing a sampling location where currents would draw the tailings water away in the opposite direction. This would essentially be a control site, not a tailings sample site, in other words. Beeton reported that his 1968 samples taken from the entire western arm of Lake Superior had nearly twice the diatom concentrations of samples taken before tailings were deposited. In other words, the first steps in eutrophication were affecting not just a local site immediately adjacent to Reserve (as the company claimed) but nearly half of the lake. Beeton pointed out to the court that waiting until Lake Superior was polluted as badly as Lake Erie might satisfy the scientific demands of Reserve for perfect knowledge, but it would not protect the lake.[50]

Robert Burd, a witness for the Minnesota Pollution Control Agency, argued that Reserve's models of assimilative capacity were deeply flawed, saying, "The Department of Interior no longer accepts the concept of the assimilative capacity of the water (adding as much as you can without ruining the uses of the water). The idea is to keep out everything you can; the reason is that in assessing assimilative capacity the scientists have been wrong too often, and we should err on the side of safety."[51] It was a significant change in conceptualizing complex systems.

Reserve's scientists continued to testify that tailings were completely inert, harmless, and fixed in place. But under cross-examination, their testimony faltered. Robert Bright, consultant for Reserve, stated that "Reserve's tailings are not being carried to any significant extent outside the western arm of Lake Superior," which is typically defined as the fifty-mile-long section of the lake from the Apostle Islands to Duluth. Bright added that "the tailings in the waters around the Reserve plant are being buried in sediment without having an appreciable effect on any aspect of the Lake." But under cross-examination, Bright admitted the opposite.[52]

Reserve's consultants argued that the mineral cummingtonite, part of the tailings, could not be used to track them because it could sometimes be found in other streams. But this core argument was undermined by Reserve's chief engineer, Kenneth Haley, who when cross-examined testified that Reserve was discharging "approximately 4.7 million tons of cummingtonite per year" into the lake. He acknowledged that the only place Reserve found cummingtonite entering the lake naturally was in a single sample from a single tributary—a finding that wasn't repeated in other samples from that same tributary. It eventually emerged that it was from contamination by road crews who had spread taconite tailings over a bridge. Reserve, it turned out, had intentionally selected this site, knowing it would undermine the state's ability to trace taconite tailings throughout the lake. Nonetheless, Haley continued to insist that tailings couldn't be mapped because "the presence of traces of cummingtonite could possibly be the result of tributary contribution."[53]

Reserve's internal memos show that they were doing their best to negate the studies showing taconite tailings spreading through the lake. One memo told Reserve scientists to focus their attention on finding cummingtonite sources inside tributaries, where Reserve tailings could not have migrated, so that these data could be used to discredit all tracer studies. The letter ordered: "A survey of the rivers for cummingtonite should be made during the spring runoff. Additional surveys should then be made of those rivers which contained cummingtonite during this survey or during the 1969 surveys. River bottoms should be cored and analyzed for cummingtonite. Lagoons associated with the rivers should also be analyzed for cummingtonite."[54]

Even Fred Lee, the Reserve consultant who had attacked Stoddard so vehemently, faltered in his testimony. Lee initially testified that tailings were harmless and had no effects whatsoever on the lake. But under cross-examination, he admitted that the tailings stimulated algae growth. He also acknowledged "that small amounts of phosphorus from tailings can stimulate algae growth" and that "oxygen demand of tailings could be quite significant," leading to poor fish conditions. Most important, he agreed that "cummingtonite is acceptable as a tracer" for tailings. Contrary to Reserve's long insistence, Lee's testimony revealed that Reserve's tailings could be tracked as they moved from the launder site into much more dispersed spaces.[55]

In an internal 1970 report for Reserve, Lee wrote that "small amounts of taconite tailings do cross the Minnesota-Wisconsin state line." It was an explosive statement—or would have been if Reserve had let the state know—because it undercut Reserve's core argument that the tailings had stayed put. But Lee went on to insist that "just because tailings are found to cross a political boundary such as a state line, this does not mean that there is presumptive evidence for a deterioration in water quality." Lee also noted (for the first time) that tailings were indeed changing the makeup of plankton in the lake. But he insisted, once again, that changes in plankton did not necessarily mean the lake was being harmed: "there is no technical basis for saying that because the number of pontoporeia or other bottom fauna organism changes by a few percent that this represents a significant effect on water quality in the lake. The overall role of bottom fauna in the ecology of a lake is poorly understood. It is unreasonable to state that changes in the bottom fauna of the order that are likely to occur as a result of Reserve's discharge would have a significant deleterious effect on water quality or ecology of Lake Superior."[56] For months, Lee and Reserve had argued that Reserve's tailings were completely inert, so they could not possibly cause changes to lake ecology. Now Lee had admitted the tailings were not inert, were crossing state lines, and were stimulating the growth of algae. But he insisted that any regulatory action required firm proof that the changes were causing the collapse of the entire lake—proof that, as a water chemist, he knew would be impossible to provide.

Privately, Lee was beginning to worry about the amphibole fibers that could indicate asbestos and had been found in Reserve's tailings. In a letter to K. M. Haley of Reserve, Lee warned that a recently published book by chemist F. C. Loughnan stated "that amphiboles rarely survive the weathering environment." This meant that any amphiboles that were found on the lake bottom came not from natural sources but from Reserve's dumped tailings. Lee added, "This statement is extremely damaging."[57]

Reserve responded to these damaging findings by calling for more research. Before the state could enforce its water standards, Reserve insisted, a comprehensive program of research that would take decades to complete should be conducted: "Statistical studies of bottom fauna should be continued. More data must be obtained from the tailings area. We must be careful in trying to explain any differences in bottom fauna populations. It was suggested that the water adjacent to bottom fauna samples be analyzed."[58]

Asbestos Controversies

By the early 1970s, the voices protesting Reserve included local environmental organizations, commercial fishermen, and sport-fishing groups — as well as federal and state agency staff. But until a connection to human health emerged, Reserve managed to delay action, always claiming more research was needed. In December 1972, Arlene Lehto, a local woman representing the environmental organization Save Lake Superior, went to a meeting of the International Joint Commission in Duluth. She told the commission that the tailings contained a form of asbestos that could cause cancer.[59] People drinking water from Lake Superior might be ingesting it, she noted, and because Duluth's water supply came directly from the lake, residents of the largest city in the basin might be exposed.

Although the commissioners and the press took no particular notice of Lehto's testimony, the federal lab staff and the Minnesota Pollution Control Agency staff did.[60] Eventually, on behalf of the brand new federal Environmental Protection Agency, the Department of Justice filed a lawsuit against Reserve in 1973, beginning a trial and appeals process that would last for a decade. The Minnesota Pollution Control Agency joined the suit, convinced by Lehto and its own scientists that Reserve was endangering public health.

In early June 1973, Judge Miles Lord heard testimony from a specialist in asbestos exposure, Dr. Irving Seikoff. Seikoff confirmed that the city's drinking water contained asbestos from the tailings, and a "thorough study

should be done on the effects of lake water on the human body." Judge
Lord initially put this testimony under an order of secrecy, fearing public
hysteria in Duluth. But on June 15, 1973, after "considerable debate in secret meetings," the public was informed that asbestos fibers, most likely
from Reserve Mining Company's tailings, had been found in Duluth's water supply. The concentration was surprisingly high: 100 billion fibers per
quart of water, which was at least "'1000 times higher' than any asbestos
level previously found in any water sample."[61] The plant's exhaust stacks,
citizens were told, were also emitting asbestiform fibers into the air.

Outrage, first local and then national, resulted. The army rumbled
into Duluth to provide clean water in huge trucks, the mayor announced
that drinking the water could kill thousands, and when the federal trial
against Reserve officially began in August 1973, the general public, "already
alarmed by reports of asbestos deaths around the country," became fixated
on the trial.[62]

The trial featured a prolonged series of scientific consultants testify-
ing on behalf of Reserve and the government, arguing over arcane details of
mineral forms, sources, mobility, and target organs. Reserve argued that its
tailings couldn't possibly contain asbestos, and that asbestos, even if it were
present in the water, couldn't possibly cause harm when ingested. Initially,
Judge Lord appeared skeptical of the prosecution's scientific claims of risk,
but after months of testimony, he ruled that there was indeed a significant
potential for health and environmental risks from the tailings. Lord first ordered the two sides to work out a negotiated settlement that would prevent
the closing of the plant, but Reserve refused. He then called directly on the
chair of the company, C. William Verity, to end the dumping of tailings.
Verity read a statement to the court "stating Reserve's waste wasn't dangerous, that it would bear no responsibility, and it would build a land dump
provided the government pay for it." It infuriated Judge Lord, who in April
1974 ordered the plant to be shut.[63]

A few key points emerge from this case. First, it was the first time
environmental health concerns had shut down the single dominant economic industry for an entire region. Second, it was essentially a precautionary move on Judge Lord's part, and he admitted as much. He knew

that the scientific evidence that those particular tailings might cause cancer was contested and uncertain, but he also refused to believe that absolute certainty was necessary before taking action.[64]

What made Judge Lord decide on the side of precaution? His actions appear to have stemmed from his growing fury that the company kept appealing to scientific reason, even as increasing evidence emerged that the company was consistently lying about scientific evidence and distorting scientific process. For example, Lord was angered when he learned "that by 1972 Armco and Republic engineers and executives had secretly developed plans demonstrating feasibility of on-land disposal [even though they had testified under oath that they could never even consider disposing of tailing on the land]." As a witness noted at the time, in Judge Lord's view, "Reserve had now forfeited any right to such consideration by repeated acts of 'bad faith'"[65]

Similarly, Dr. Donald Mount of the EPA provided evidence that Reserve was distorting scientific process. The company designed studies with false controls to contaminate any possible findings, and it took measurements of tailings in places where it knew could not possibly show significant differences from controls.[66]

Bad faith became a continued refrain as judge after judge confronted Reserve's intransigence—so much so that the court instituted formal "bad faith" proceedings against Reserve. On May 11, 1975, Judge Lord summarized the evidence that Reserve had lied and suppressed evidence, stating: "After listening to testimony for over nine months, the court has formed the opinion that the credibility of the defendants collectively in this case is seriously lacking. They have misrepresented matters to the court, they have produced studies and reports with obvious built-in bias, they have been particularly evasive when officers and agents were cross-examined." A year later, Judge Edward Devitt noted, "Reserve's bad faith in the conduct of the defense of this lawsuit and its failure to truthfully comply with discovery requests and court orders justifies sanctions. . . . The concealment of the materials was part of a course of conduct adopted by Reserve in bad faith 'for the sole purpose of delaying the final resolution of the controversy.'"[67]

Judge Lord became particularly infuriated that the company kept playing the "you'll destroy jobs!" card to justify continued pollution. He stated: "In essence, defendants are using the work force as hostages. In order to free the work force at Reserve, the court must permit the continued exposure of [the citizens of Duluth, Minnesota, and other North Shore communities] to known human carcinogens. The court will have no part of this form of economic blackmail."[68]

Lord eventually found Reserve guilty and ordered the beneficiation plant to close until the company had constructed a land-based tailings site. While conservationists were pleased, Edward Davis, the engineer who had promoted taconite, was horrified that, in his view, government regulators could "wield such tenuous evidence to foster widespread alarm and potentially cut off the economic lifeblood of an entire region." The appeals court seems to have agreed, for it quickly reversed Lord's closure of the plant, stating that nobody had proved beyond doubt that citizens would die. Russell W. Peterson, chair of the White House Council on Environmental Quality, attacked this decision, warning that virtually the only way anyone could prove a case involving health hazards is by "counting dead bodies through an after-the-fact epidemiology study." Although a federal appeals court did allow Reserve to continue dumping in the lake for six more years, eventually the company was forced to relocate its tailings disposal to a land-based site. Thomas Huffman writes that it was "a landmark decision, one that gave the EPA broader powers to regulate corporate pollution, a practice unheard of before the lawsuit."[69]

Why did decades pass before Reserve's pollution was regulated? Not because of a lack of concern: beginning well before the plant opened, numerous citizens protested that pollution. The spatial and historical context of toxic contamination made it appear to regulators that potential risks would be small. Because many earlier iron mines had been underground hematite mines, which exposed few contaminants directly to air and water, the historical experience may have shaped the ways that scientists and regulators approached the potential for mobilization of toxics from new mines. The key stage for regulators who were trying to use scientific knowledge to balance risks with benefits was when new scientific information emerged that questioned the safety of an established project. Typically, each time regulators reached the limits of their knowledge about the potential risks of contaminant exposure, they decided to move ahead with exposing people and environments to toxics. Each time they vowed to use exposure as an experiment of sorts, one that would be monitored and learned from. Contaminants were released with the underlying assumption that any major problems would emerge in time for corrective action. How well were regulators actually able to accomplish this? Not very well, to put it mildly.

Only after claims that asbestos had been mobilized from taconite disposal into the drinking water and bodies of urban residents distant from the disposal site did the federal and state governments officially begin to question risks from taconite. The plant's exhaust stacks were also emitting asbestiform fibers into the air. Reserve Mining Company disputed evidence about toxic mobilization, arguing in complex ways that asbestos could not possibly move from tailing into Lake Superior, from Lake Superior into drinking water, and from drinking water into lungs. Moreover, data suggesting toxicity did not always make it out of industry labs into regulatory spaces. Michael Egan, in his essay on the coproduction of environmental knowledge and mercury regulation, discusses one Swedish chemist who intentionally "obfuscated the debate . . . [by suggesting] that the high mercury levels might be attributed to naturally occurring mercury in Swedish rocks and soil, as well as industrial mercury."[70] Similar processes occurred in taconite mining debates. Reserve Mining Company suggested that pollution was due to natural sediments, not their industrial waste, while also suppressing scientific information that would damage its case. Reserve also designed studies that would hide rather than reveal the movement of asbestos fibers from taconite tailings.

By the time of the federal court case against Reserve, the 1972 amendments to the Federal Water Pollution Control Act (known as the Clean Water Act) had finally transferred some of the authority to regulate water pollution to the federal government. With these amendments, historian of science Stephen Bocking writes, American "cooperative pragmatism had been replaced by a more legalistic, adversarial approach to regulation. A wider range of interest groups now participated in forming policy, in the courts and in other arenas. Activists demanded that the federal government take the initiative away from the states, viewing it as less likely to be influenced by local economic interests."[71]

When Reserve's pollution mobilized from a local site into broader waters, it also mobilized regulatory power from the local spaces of the state into broader federal spheres. Concerns about the mobility and toxicity of Reserve's tailings mushroomed during the 1950s through the 1970s, even as taconite boosted economic development in a struggling region. By the 1970s, faith in state committees of conservation had crumbled as citizens and scientists confronted new evidence of corporate corruption and chemical toxicity.

Mining, Toxics, and Environmental Justice for the Anishinaabe

AS NOTED IN EARLIER CHAPTERS, mining wastes mobilize toxics into broader watersheds. Because those toxics accumulate in ecosystems and bodies, their legacies can be remarkably persistent. New technologies and methods for the detection of toxic chemicals have drawn increasing attention toward the pervasive presence of persistent, bioaccumulative toxic contaminants. Some of these, such as PCBs, are legacy contaminants — long banned but still sticking around. Others are emerging contaminants, not yet banned or fully understood. Many are synthetic compounds, and others are natural chemicals that have been unnaturally concentrated and mobilized by mining.

Toxic chemicals released by mining move from their sites of production and consumption into much broader and dispersed spaces. As they flow into water, they bioaccumulate in fish and eventually make their way into the people who eat that fish. These contaminants have permeated global ecosystems, crossing international boundaries to affect people far from initial sources of production and consumption. Their toxic residues not only complicate political boundaries, they also confuse temporal distinctions, for their legacies persist long after they have been banned. Moreover, the risk of exposure to these chemicals is rarely equitably distributed in a human population.

Unlike gold or nickel mines, it is possible for iron mines to release relatively few toxic wastes into the larger watershed and atmosphere. Some, but not all, iron mines release sulfides into the water, which create acidic drainage issues that mobilize toxics into the watershed. Some, but not all, iron mines include processing facilities that release mercury into the atmosphere, where it then mobilizes into the larger environment as methylmercury and bioaccumulates in fish. Some, but not all, iron mine tailings destroy local wetlands and aquatic habitat. Whether or not a particular iron mine has these consequences depends partly on the geologic context, in particular the composition of the overburden (the rocks that cover the ore deposit). However, it's not the geology alone that leads to toxic consequences; social and political decisions about how to interact with that geology ultimately shape toxic outcomes. In the Lake Superior basin, conflicts over Indigenous land tenure rights and past toxic exposures continue to play central roles in current mining controversies.

Beginning in 2010, the Lake Superior basin witnessed a new mining boom that rivaled the post–World War II boom, with dozens of companies actively exploring the region (figure 6.1). State and provincial governments

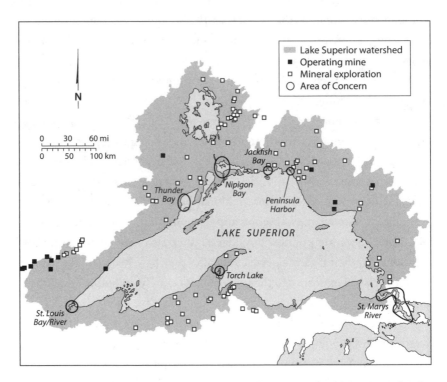

Figure 6.1 Map showing extent of mining exploration and activity around Lake Superior, 2015. (Drawn by Bill Nelson, modified from map created by Great Lakes Indian Fish and Wildlife Commission.)

supported the mining companies, just as they had in the post–World War II era, for the companies promised new revenues to economically depressed northern regions with few steady industrial jobs.

In 2011, a company named Gogebic Taconite (GTAC) formed in order to develop the largest open-pit mine in the world—just upstream of the Bad River Band's reservation on Lake Superior. Owned by Cline Resources Development (a company largely focused on coal), GTAC announced that, even without experience in iron mining, it would mine and process Wisconsin's taconite ore body to take advantage of Asia's building and steel commodities boom.

The mine would have been sited just upstream of the reservation boundary, and the waters flowing out of the mine site would have contaminated water, fish, and Indigenous communities living downstream (figure 6.2).[1] If permits had been approved, the GTAC mine would have been located within ceded territories, the locations where the tribes retained hunting, fishing, gathering, and comanagement rights when they signed

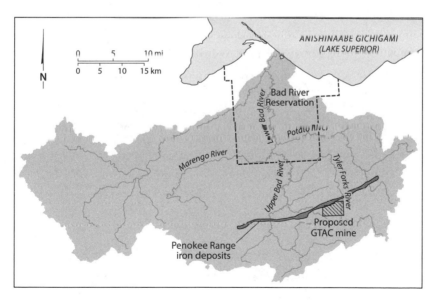

Figure 6.2 Map of taconite ore body in relation to the Bad River Reservation. (Drawn by Bill Nelson.)

the nineteenth-century treaties enabling white settlement (figure 6.3). Bad River Band members point out that it is no accident that the proposed mine site was within ceded territories in their watershed. As environmental justice scholars argue, environmental exposures vary in their spatial distribution, and social inequalities influence vulnerability. Tribal members have often borne the greatest burden from the toxic wastes fostered by past mining projects, but they have rarely had much decision-making power in the planning process.[2]

The Bad River runs through the potential mine site before entering the 16,000-acre Kakagon–Bad River Sloughs, the largest undeveloped wetland complex in the upper Great Lakes. In 2012, it was designated as a Ramsar

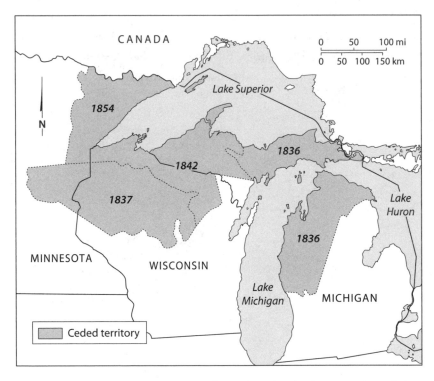

Figure 6.3 Ceded territories, showing dates of treaties, in the upper Great Lakes. The Lake Superior tribes signed treaties in 1836, 1837, 1842, and 1854 with the United States government, ceding land but retaining the rights to hunt, fish, and gather in the ceded territories. (Drawn by Bill Nelson.)

site, recognized by the United States as a wetland of international importance. The United Nations convention noted that "as the only remaining
extensive coastal wild rice bed in the Great Lakes region, it is critical to
ensuring the genetic diversity of Lake Superior wild rice."[3] The sloughs
make up 40 percent of the remaining wetlands on Lake Superior's coast,
and they contain the largest natural wild rice beds in the entire world.

For members of the Bad River Band, these wild rice beds are central
to their cultural identity and economy. *Manoomin,* the Anishinaabe word
for wild rice, translates as "the good berry," yet manoomin is more than
food — it is a sacred gift from the Creator. According to oral tradition, the
Anishinaabe left their homes along the Atlantic Seaboard, perhaps about fifteen hundred years ago, and journeyed into the Upper Great Lakes. Their
stopping places included Niagara Falls and Sault Ste. Marie at the outlet of
Lake Superior, where they found excellent fishing. At the Straits of Macki-
nac, they split into three groups: the Potawatomi, who moved south into
the area between Lake Michigan and Lake Huron; the Ottawa, who moved
north of Lake Huron; and the Ojibwe, who explored the south shores of
Lake Superior. Upon reaching the St. Louis River estuary at the base of
the lake (near today's Duluth), the Ojibwe found manoomin, the "food that
grows on water." Fifty miles east, they found their final stopping place: an
island they called *Mooningwanekaaning* (now known as Madeline Island).
Madeline Island became the economic center of Anishinaabe trading (and
site of a key fur-trading post centuries later) and its spiritual center as well.
As Patty Loew writes: "The Three Fires, as the Anishinaabe referred to
their religious alliance, returned to the island at various times of the year
to conduct *Midewiwin,* or Grand Medicine Lodge ceremonies. The Midewiwin, through its songs, stories, and rituals, embodied the spiritual heritage of the Anishinaabe and offered a code of conduct to keep them culturally rooted and physically and spiritually healthy."[4]

Ancestors of the Bad River Band chose to make their homes along
the Kakagon Sloughs just a few miles south of Madeline Island because
the wild rice beds they found there had figured significantly in their vi-
sions. In the Kakagon Sloughs, the Ojibwe found manoomin in abundance.
Manoomin became a major portion of Ojibwe subsistence, for it could be

stored throughout much of the winter, providing sustenance during lean times. Wild rice, however, is sensitive to hydrological changes from development and to sulfates in the watershed. Because some taconite mines can mobilize sulfates into water, band members argued that stopping the mine was essential for their survival. Pollution would threaten wild rice beds and fisheries, and, even more important, it could fracture cultural and spiritual connections with the past.

In contrast, many Euro-American residents of Hurley, Wisconsin, and other nearby communities argued that the proposed mine was the only thing that could rescue them from the economic devastation that followed the closure of local hematite iron mines in the 1960s. Hurley lies just outside of the watershed that would be affected by the proposed mine, thus the town would not be directly influenced by any water quality issues presented by the mine. Denying permits for the proposed mine, Hurley residents argued, would amount to a form of economic suicide.

As John Sandlos and Arn Keeling have argued for Pine Point, Canada, "complex and contested meanings of place and community" are common at mining sites. "While regarded by 'outsiders' as brutal, degraded or even toxic, former mining landscapes may be touchstones of community identity and memory and provide both material and cultural resources for economic recovery or even political resistance."[5] This is certainly true for iron mines in the Lake Superior basin. Different communities within the basin have quite different interpretations of their mining past, and these views about the past help to shape their perspectives on current mining proposals.

The Penokee Range lies in what is now northern Wisconsin; across the border in the Upper Peninsula of Michigan, the same range is named the Gogebic Range. In this chapter I use the term "Penokee-Gogebic Range" when referring to the entire range. Once the center of a thriving— but short-lived—hematite mining economy followed by clear-cutting, the region's forests have now grown back (figure 6.4).

The first phase of the proposed GTAC taconite mine would have created a pit five miles long and one thousand feet deep. Eventually, plans called for twenty-two miles along the ridge of the Penokees to be carved off.

Figure 6.4 Aurora Iron Mines, Ironwood, Michigan, between 1880 and 1899. The hematite iron mines in the Penokee-Gogebic Range had intense but short-lived effects on the forests and watershed. (Detroit Publishing Company. Library of Congress LC-DIG-det-4a04172.)

The proposed techniques bore little resemblance to the deep-shaft mining for hematite that was the basis of the region's iron mining history. Rather, it would become the first "mountain top removal" mine in the upper Great Lakes region. As one anti-mining activist wrote: "Gogebic Taconite . . . has grandiose plans: To blow them to smithereens with a series of 5.5 million-pound explosives—each similar to the impact of the bomb dropped on Hiroshima—in order to extract low-grade iron known as taconite. Waste rock with the potential to leach billions of gallons of sulfuric acid from what would be the largest open pit iron mine in the world could be dumped."[6] While the rhetoric here was exaggerated, the details were not.

The proposed mine would have targeted a band of iron-rich ore in the Penokee-Gogebic Range known as the Ironwood Formation. Best estimates suggest that the Ironwood Formation contains at least 3.7 billion tons of taconite ore that could be mined economically, or 20 percent of known iron ore deposits in the United States. This translates into 1 billion tons of steel, or sixty-six years of domestic supply, making it a significant ore deposit in the national and indeed global context.[7]

The Ironwood Formation is such an enormous deposit, GTAC argued, that it was inevitable that it would be mined. The language of inevitability figured heavily into the pro-mining discussions. But, as U.S. Steel found when it did bulk sampling in the 1980s and decided against mining taconite there, the geological context of the Ironwood Formation made it extremely difficult to exploit without losing money.[8] The rock is very hard (requiring enormous blasting capabilities), and the deposit is tilted at a 65-degree angle, overlaid with 650–985 feet of overburden and banded with quartzite and shale—all details that require extensive energy and infrastructure for extracting ore economically.

Environmental regulations for mines were more relaxed in the 1980s, and GTAC had never mined taconite. How then could GTAC propose to mine such a deposit, when U.S. Steel had decided it was impossible three decades earlier?[9] The simplest ways to mine taconite cheaply are to reduce labor and environmental costs. And that is what GTAC decided to do in 2011, when rising steel prices in Asia made investments in new mines seem attractive. New, enormous mining machines had been developed that could extract up to two hundred tons of rock out of the mine in a single load, thus lowering labor costs. Reducing environmental costs would be possible if the industry could block the implementation of new federal standards to limit mercury emissions from facilities processing taconite ore. And, if a company could persuade a state to exempt the industry from water quality regulations, tailings could be dumped cheaply into streams and wetlands.

Some environmental groups opposed to the GTAC mine argued that the Lake Superior basin and the Penokee-Gogebic Range in particular are essentially pristine and should never be mined. Residents of former iron-mining communities argued just the opposite: they said it has long been an iron-mining area, so it should continue to be one. Why did different groups have such different interpretations of the past, and how did these contested stories affect policy disputes? When the Anishinaabe were forced onto reservations in 1842 to make the rest of the region available to miners, disease, poverty, and despair often resulted. Yet the Anishinaabe successfully defended their right to remain on their homeland and its waters. They retained usufruct rights to the ceded territories, making certain that their members could continue to hunt, fish, and gather in perpetuity. These treaty rights have become central to the governance of mining conflicts in recent decades.

Mining in the Lake Superior basin is not new, but the technologies for extensive extraction are recent. An 1848 report by A. Randall described the presence of hematite iron ore in the Penokee-Gogebic Range, and extraction began in 1886. To the west in Minnesota, on the Mesabi Range, iron mining began with discovery of hematite iron ore in 1865, production in 1885, and rapid expansion through the 1890s. In both places, mining efforts targeted the high-grade hematite ores that were concentrated and did not require extensive processing before being shipped across the Great Lakes to steel mills. Forty underground mines worked Wisconsin's Ironwood Formation between 1877 and 1967, extracting 325 million tons, which were shipped to blast furnaces in the lower Great Lakes.[10] Mining towns such as Hurley, Ironwood, Iron River, and Montreal thrived during the hematite era. When the brief boom collapsed, first alder then maple trees grew back, hiding the shaft holes and cloaking some of the slag piles, allowing people to imagine the forests as pristine and untouched.

During the mine-permitting process from 2011 until 2014, GTAC tried to present taconite as a safe ore to mine with minimal toxicity concerns. It is so safe and so pure, GTAC insisted to the Wisconsin legislature, that the state should quickly pass a new law exempting taconite mining from environmental regulations. The argument had two parts: first, because taconite itself rarely contains iron sulfides or pyrites—which can produce acid mine drainage—it is pure. Second, the extraction process uses magnets and clean water, not hydrosulfuric acid.[11] Both arguments were familiar from the 1947 permitting debates over the Reserve mine.

Technically, both arguments are correct, but the conclusion that follows (therefore taconite mining has no toxicity concerns) ignores the larger geological and biological contexts of taconite within a watershed. First, taconite is a low-concentration iron ore, and to extract the valuable part, the rest of the rock (the tailings) must be crushed to a fine dust, mixed with water, then dewatered and stacked in piles or dumped into water. As the Reserve Mining episode detailed, taconite tailings are so fine that they are easily mobilized into much more dispersed spaces. At least 70 percent of the volume of the Ironwood Formation would be turned into waste and fine tailings, and the waste for the initial four-mile stage of mining alone would create a pile "600 million cubic yards" in size, stretching "one mile square, 600 feet high."[12]

Second, there is the problem of those 650–985 feet of so-called over-burden: the rock on top of the taconite itself, and the world on top of that rock. The very term "overburden" suggests that a living ecosystem—for-ests, forbs, birds, habitat, streams, the many different communities within the soil, the layers of rock that lie under the soil—is nothing but a burden that blocks the true resource, taconite. The tribe has resisted the terminol-ogy of mining, arguing that it renders invisible the biological, geological, and hydrological connections that sustain them.[13]

Asbestos in tailings continues to present an enormous environmental challenge, just as it did for Reserve Mining Company in the late twenti-eth century. GTAC had promised that they would avoid earlier problems with taconite tailings by dewatering them and stacking them on land, in huge piles called "dry stacks." But tailings particles in dry stacks are easily eroded by wind and water, and mobilized into the watershed. Asbestiform fibers are released from taconite processing when gruniform minerals are present (as they were in the Reserve Mining situation). Exposure to these asbestiform fibers has been linked to threefold increases in mesothelioma in the Minnesota Iron Range close to existing taconite mines.

How do we know if GTAC's taconite tailings might contain asbestos? Most taconite deposits do not, and within the Lake Superior basin, there is enormous variability. A detailed review of taconite health risks published by the Wisconsin Department of Natural Resources in December 2013, however, suggests reasons for concern. Work by the geologists Tom Fitz and Marcia Bjornerud, who both examined the ore from old bulk sampling sites near the proposed mine, confirmed the presence of asbestiform fibers in site of the proposed GTAC project.[14]

Yet these findings remain contested. The corporate owners of the land initially denied any asbestiform fibers and refused to allow geologists access to their samples. After Fitz's and Bjornerud's results were reported, the Ashland County board chair, Peter Russo, said "he remains doubtful a mine could be built. . . . Until they can show me that they can mine that material safely, there's no way in God's green Earth I'd say 'do it.'" Russo added that "health and environmental risks appear to outweigh any economic gain from mining the Penokees." In the late fall of 2015, the corporate owners of the land agreed to move forward with a study of the core samples, in partnership with Congdon Mineral Management, a firm associated with La Pointe Iron, one of the landowners. They partnered with retired United States Geological Survey (USGS) geologist Bill Cannon and three geologists from the Forest Service national headquarters to study samples from exploration drill cores taken years before. The USGS "signed a confidentiality agreement in order to gain access to the core samples," meaning that no "preliminary findings will be released without the consent" of the lease owners.[15] However, the agreements stipulate that the results may eventually be published. USGS-generated data cannot be suppressed, according to the terms of the agreement, which will allow stakeholders in the region to better understand the risks of sulfide-bearing minerals.

Discussions of new taconite mines continue in the shadow of the Reserve Mining Company's release of tailings into Lake Superior and the resultant movement of asbestos into Duluth's water supply. But interpretations of this history differ sharply. When GTAC argues that taconite is perfectly safe, they interpret the Reserve Mining history as an example of what happens when environmentalists and regulators overreach and shut down a region's economy for an unproven threat. While epidemiological studies have shown that miners on the Mesabi Range have a 300 percent higher rate of mesothelioma from airborne asbestos exposure than the general public, no one has shown that people in Duluth died at higher rates from waterborne asbestos, which has been shown in other studies to be associated with higher rates of gastrointestinal cancers (including stomach, colon, and esophagus). Of course, no one has shown that they didn't die at higher rates either; the funding for the long-term studies that were promised back in the 1970s dwindled in the 1980s. Doses of a billion asbestos fibers per quart have been shown to cause gastrointestinal tract cancer, and some of Duluth's drinking water during Reserve's operation had that level of contamination. But because many of these cancers have a forty-year latency period, without funding to follow the specific people exposed in Duluth we simply do not know. However, studies published in 1997 showed that women living in the region had an excess of cancer compared to women living in the Twin Cities, with 30 percent higher rates of esophagus and stomach cancers.[16]

For many environmentalists in the basin, the Reserve Mining history suggests how treacherous taconite mining may be to water quality, wild rice, and human health. Water in the Lake Superior basin is an increasingly vulnerable and critical resource. You might think that with the biggest lake in the world, there would be plenty of water to go around, but the watershed itself presents challenges for responsible mining. Ore mined and processed from the initial four-mile-long GTAC pit would require five to ten trillion gallons of water for the processing alone. With a pond three feet deep, the water used in processing the taconite would cover eighty-nine square miles. Mines also need to be dewatered when the hole dips beneath the groundwater level; otherwise the pit quickly fills with water. One report notes: "Pumping would certainly draw down the water table in the area, so wells close to the mine would have less water than today, or even dry up completely. The hydrogeology is not well understood though, so the extent of impacts on groundwater—like many of the environmental impacts that come with mining—are uncertain."[17]

The Bad River Band feared that such intense groundwater pumping would devastate their springs and the Kakagon Sloughs just downstream. For the tribe, degradation of the ecosystem would be devastating. "Wild rice is central to Ojibwe culture and lifestyle and the Bad River Ojibwe Tribe has worked hard to protect the wetlands ecosystem. As a sovereign nation, the tribe has legal authority to regulate water quality of the Bad

River. Their water quality standards will influence what can and will happen in the headwaters of the river."[18]

In addition, the water that flows through the mine or tailings and continues downstream raises concerns about acid drainage—one of the most persistent risks from mining. Acid mine drainage represents a mixture of natural and constructed toxicities. Many iron formations contain heavy metals that would be toxic if they were mobilized into biological systems. Typically, they are bound in stable formations, where they don't move into the atmosphere or the water supply on a time scale that matters for biological life. (Over millions of years, they do mobilize, suggesting how important considerations of scale will be for this project.) But when acid conditions are present, those chemicals and heavy metals can rapidly move into biological systems. Mines with acid drainage issues will need to be cared for in perpetuity, guarding against the acidity and the toxic leakages that can flow for millennia, altering ecosystems, eradicating wild rice and the cultures that depend on it.

The presence of pyritic ores in the formation has been denied by GTAC (and the company has refused to allow the state, the tribes, or local citizens to view the samples it obtained from U.S. Steel). However, in 1929 the Wisconsin Geological Survey reported that pyrite was associated with local ore and waste rock, and a 2009 USGS report came to the same conclusion. When ground to a fine dust (as required for taconite extraction) then exposed to oxygen and water, pyrites create sulfuric acid, creating acid drainage that leaches such harmful chemicals as lead, arsenic, and mercury into groundwater and surface water. Selenium, mercury, arsenic—perfectly natural chemicals—lie bound and buried in rocks until miners release them while digging for something else that has become defined as a resource. Then as waters move through mining sites, these chemicals move into fish bodies and from there into human bodies.

Nancy Schuldt of the Fond du Lac Band describes the "fingerprint" of Minnesota's mines on water quality. Sulfates are created when rock formations are blasted or crushed and sulfur in the rocks comes in contact with air and water. One major issue is that taconite tailing disposal basins are designed to seep waters into downstream creeks, in order to keep the disposal basins from overflowing during heavy rain. The effluent from the seepage sometimes contains elevated levels of sulfate, which impair wild rice beds, increase methylmercury in waters and fish, cause eutrophication of lakes and rivers, and at some sites kill aquatic species and communities. Taconite plants have also been required to install scrubbers on smoke stacks so that sulfur compounds are not released into the air. Yet the highly concentrated scrubber wash water is emptied into tailings basins, adding to the toxicity of anything that leaches from them.[19]

Some taconite mines in Minnesota have continued to leach sulfates downstream decades after closure. Historical studies have shown that wild rice was abundant in the upper St. Louis River watershed (above Duluth) before the 1950s, when taconite mining boomed. Currently, sulfate levels are high in the St. Louis River, and wild rice stands are few and stunted. The taconite tailings basin once owned by Ling-Temco-Vought (LTV Steel) still leaches sulfates and other contaminants into the St. Louis River, and from there into Lake Superior. Elsewhere on the north shore, a tailings basin owned by Minntac is leaching three million gallons per day of sulfates and related pollutants into two watersheds.

The Dunka mine, a taconite mine near Babbitt, Minnesota, was covered with sulfide rock similar to the overburden present in the Penokees. Its history suggests some of the difficulties of containing pyritic materials. The Dunka mine was operated by LTV Steel from 1964 to 1994, making a twenty-million-ton waste rock pile that was a mile long and eighty feet high. This waste began leaching copper, nickel, and other metals into wetlands and streams almost immediately. Decades later, an average of 300,000 to 500,000 gallons continues to run off them each month, according to Minnesota Pollution Control Agency (MPCA) documents. Between 2005 and 2010 the runoff had violated state water standards nearly three hundred times, yet rather than force the company to spend the money to stop the toxic runoff, the MPCA fined the company that now owns the site (Duluth Metals Limited) $58,000 — a cost of doing business that is far cheaper for the company than cleaning up the tailings. Yet the contaminated water flows into the Boundary Waters Canoe Area Wilderness, an area that is supposed to be protected from toxic discharges.[20]

What can these histories of mine problems tell communities that are trying to decide about the potential for acid drainage from proposed new mines? Christopher Dundas, chair of Duluth Metals Limited, argues that historical problems have no bearing on or relevance to the future proposed mines. "This is a completely different era than what happened in the '60s," said Dundas. "Our operation will be state of the art and will be totally planned and designed to absolutely minimize every environmental issue." History is irrelevant, in other words. But to advocates for the Boundary Waters, history matters. One opponent fears that problems with the Dunka mine are "an indicator of problems to come" from proposed mines. A former state employee, Bruce Johnson, who regulated Dunka and other local mining issues, told reporters that "he fears that state agencies will shortcut environmental rules because of the intense political pressure to approve mines and put people to work. 'I want to have good jobs, too, but I want to do it right,' Johnson said. 'These guys are going to make multi-millions of dollars. We don't want to be left with a bunch of mining pits full of polluted water that even ducks won't land on.'"[21]

Wild rice is particularly sensitive to even extremely low levels of acidic drainage, creating enormous concerns for the tribes. Past taconite mines in Minnesota have continued to leach sulfates into wild rice beds decades after closure. The tailings basin once owned by LTV Steel still leaches sulfates and other contaminates into the St. Louis River, and from there into Lake Superior. Elsewhere on the north shore, the Minntac taconite tailings basin is leaching three million gallons per day of sulfates and related pollutants into two watersheds. Minnesota does have a sulfate standard for taconite facilities that requires facilities with wild rice downstream to limit their sulfate discharges to less than 8–10 parts per million. That standard was adopted in 1973, but the Minnesota Pollution Control Agency only once tried to apply it to a taconite permit. The industry facility immediately sued the MPCA, and the agency halted environmental enforcement. Right now not a single facility on the range is in compliance with the sulfate standard (which is currently being reworked). Nancy Schuldt of the Fond du Lac Band said, "Recently, the Environmental Protection Agency (EPA) has begun working with Minnesota regulators to encourage enforcement, but even today, no facility on the Iron Range meets the wild rice sulfate standard." Technology for water treatment does exist, but it is expensive, and taconite facilities have not been required to treat the effluents from their basins. When asked why this standard has not been enforced, Schuldt replied: "The mining industry in Minnesota is very powerful and exerts political pressure to prevent agencies from strictly implementing the laws." She added that based on her experience in Minnesota, "agencies have great difficulty enforcing regulations that exist, and that in some instances, regulations have been weakened so that permits are not actually protecting environmental quality."[22]

Some Minnesota legislators are pressuring the Minnesota Pollution Control Agency to relax the standard rather than enforce it. For example, in 2011 the Minnesota legislature proposed an amendment to a finance bill that would have increased the sulfate limit to 250 mg/l (250 parts per million). Legislators argued that historical records are "incomplete science" and true scientific models suggested high levels of sulfates are safe—for adults drinking the water.[23] An unwillingness to interpret the evidence of historical change suggests that the boundaries between regulation, ecological history, and the mining industry remain contested.

Finally, mercury from taconite mining is a growing concern. Mercury is released from the emissions stacks during taconite processing, and taconite mining is now the primary source of mercury produced within the Lake Superior basin, surpassing deposition from coal plants. Could a new taconite plant, such as GTAC, be an important source of mercury emissions? The answer is unclear. Taconite ore varies considerably in mercury content. For example, Northshore Mining is one of the larger pellet producers but has a low level of mercury emissions (just over five pounds in 2010). So, the emissions from a new or expanded taconite plant don't depend on production levels—it is the ore itself that matters more. But because the contents of the GTAC ore samples remain hidden to the public and to regulators, it is impossible to predict mercury release.

Mercury offers an excellent example of the complex relationships between ecological change, industrial development, and contamination data. In 2009, the U.S. Geological Survey reported that mercury contamination was found in every fish tested at nearly three hundred streams across the country.[24] The highest levels of mercury were detected in some places most remote from industrial activity. Remoteness offers no protection, and the very richness of the remote wetlands increases their vulnerability to toxic conversions. Methylmercury finds its way into fish and eventually into the people eating that fish. Eating fish is of great cultural significance, particularly for Indigenous communities in the basin. But its potential contamination forces communities to make trade-offs between their beliefs and possible harm to themselves. How much fish do you eat when it's culturally important? How much do you eat when you're pregnant? These are difficult dilemmas posed by changes in watershed health. Contaminants transform more than the health of lakes, fish, and forests; they transform cultural identities as well. Interpreting the historic evidence of fish and human contamination has become a politically and culturally complex exercise.

Minnesota's taconite plants add eight hundred pounds per year of mercury to the basin, and a recent study found that 10 percent of newborn babies in the Lake Superior basin have mercury levels over EPA standards.[25] However, much of the mercury that actually *accumulates* within the basin comes not from local sources of production but from global sources, such as coal burning, that have been transported via the atmosphere into the basin. It creates a tension: Will controls of mercury emitted from local taconite processing actually reduce exposure to contaminants? Or will those controls indirectly increase exposures if they shut down the local taconite industry, displacing production to China, which might then release greater levels of mercury emissions that return to Lake Superior waters, fish, and people? Pollution is now global. Unraveling local versus global sources and exposures presents enormous challenges for regulatory communities.

Mining, microbial ecology, and mercury interrelate in complex ways in the watershed. When mining exposes natural metal sulfides in ore bodies to air and water, oxidation results, leading to acid drainage. Microbes exist in many rocks, but usually in low numbers because lack of water and oxygen keeps them from reproducing. However, during the disturbance from mining, those microbes are exposed to water and oxygen and their

numbers multiply, forming colonies that can greatly accelerate the acidi-
fication processes. These sulfates also encourage conversion of elemental mercury (not particularly toxic) to methylmercury (extremely lethal),
which then bioaccumulates in fish tissue and from there makes its way into
wildlife and people.

Mercury reductions are certainly possible from mining. Over ten
years, between 1990 and 2000, mercury associated with intentional use in
products (such as thermometers) decreased significantly in the basin, from
780 pounds in 1990 to 37 pounds in 2000. That's a 95 percent reduction. In
the mining sector, mercury also dropped in the same ten-year period—by
76 percent. But those reductions came not from improved technologies but
rather from simple closures of particularly polluting processing facilities.
When the iron sintering plant in Wawa, Ontario, closed, mercury levels
began to drop, and when the copper smelter at White Pine closed, levels
dropped even more. Each of these facilities had put out about 1,300 pounds
of mercury per year. Both were operating in 1990 but had closed by 2000.
Reductions in mercury levels are certainly possible in the mining sector.
Mercury-control technologies are available, but they can be expensive and
energy intensive (and if the energy to run those control technologies comes
from coal, mercury is emitted as well). The state of Minnesota has continued to research mercury-reduction technologies for taconite processing
(decades after the Great Lakes Water Quality Agreements agreed that all
parties in the basin would stop adding mercury in any form).[26]

The Anishinaabe and Mining on Ceded Territories

The MPCA has largely agreed to delay enforcing existing regulations on
sulfate and mercury releases from taconite plants while additional research
continues. This position frustrates the tribes, because on ceded territories
the tribes have rights to comanagement of land and water resources. The
federal government made three major land cession treaties with the Anishinaabe in the Lake Superior basin, which established reservations that were
to be exclusively under the control of the tribes. Equally important, the
tribes were careful to retain the right to hunt, fish, and gather on ceded ter-

ritories, which also meant the right to participate in management of natural resources.[27]

State governments rarely recognized these ceded territory rights, until a series of brutal assaults took place in the late twentieth century. For decades, Wisconsin arrested Anishinaabe who fished and hunted on ceded territories without state licenses, which the tribes insisted were unnecessary. In 1974, two members of the Lac Courte Oreilles Band were arrested for spearing fish on ceded territories. The Lac Courte Oreilles sued the state for treaty right violation, and in 1983, the federal court upheld off-reservation treaty rights in a landmark judgment known as the Voight Decision. The state of Wisconsin appealed (and eventually lost). Meanwhile, white supremacist vigilantes (including members of the Hurley chapter of the Ku Klux Klan) attacked Anishinaabe spear fishers who were exercising their treaty rights, and a series of violent protests at fish landings marked the late 1980s. For all the violence, as geographer Zoltan Grossman argues, in the process of fighting each other, the tribes and the whites created a series of connections that the tribes were able to build upon a few years later when Exxon proposed to build a copper and zinc mine within the Wolf River watershed (a National Wild and Scenic River). Like the GTAC mine, the Exxon mine would be located within ceded territories just upstream of an Anishinaabe reservation (the Mole Lake Sokaogon Reservation).[28] The state of Wisconsin pushed for the mine, believing it would bring jobs to the north and money to the state. The Wisconsin Department of Natural Resources decided to set standards for water leaving the mine at industrial water quality levels, allowing for forty million tons of tailings and acidic mining waste that would have destroyed wild rice beds on the reservation. To stop the Crandon mine, the Mole Lake Sokaogon worked with the EPA to win the right to establish their own clean water standards.

Tribal lawyer Glenn Reynolds wrote about the case:

Wisconsin challenged the tribe's authority to enact tribal water quality standards on the grounds that the federal government had already given Wisconsin primary authority

over the state's water resources and could not rescind that
authority and pass it on to tribal governments. Ironically,
Wisconsin argued that the Public Trust Doctrine granted the
state the exclusive right to regulate, and potentially degrade,
the water quality of Rice Lake on behalf of Wisconsin citizens.
Unsurprisingly, the mining company supported Wisconsin's
stance. Three downstream towns and a village, however, filed
a brief in support of the Sokaogon standards. After six years
of litigation, the U.S. Supreme Court declined to review a
federal appeals court decision that upheld the authority of the
Sokaogon to set water standards necessary to protect reservation waters.[29]

Tribal efforts resulted in Wisconsin Act 171 passed in 1997, which became
known as the "mining moratorium." It required a moratorium on issuance
of permits for mining of sulfide ore bodies until companies provided historical information proving that they had successfully controlled mining
waste from other mines for at least ten years.[30]

Another key instance of mining activism among the Wisconsin tribes
was the Bad River Band's 1996 blockage of railroad tracks that would
have brought sulfuric acid to a copper mine in White Pine, Michigan. On
July 22, 1996, tribal members blockaded the railroad tracks that crossed
their reservation, stopping a train headed for the White Pine copper mine
in Upper Michigan. The train was carrying sulfuric acid, for the mining
company planned to experiment with "solution-mining," which would
involve injecting 550 million gallons of acid into the mine to extract any
remaining copper. Environmentalists and tribal members argued that spills
might contaminate groundwater and Lake Superior, but the EPA granted
permission without first requiring a hearing or environmental impact statement. Anishinaabe activists insisted the project was illegal because the EPA
had failed to consult with affected Indian tribes, as required by law. After the EPA backed down and agreed to require an environmental impact
statement, the company withdrew its mining-permit application, and the
White Pine copper mine and smelter shut down.[31]

After January 2011, when news of the GTAC's proposed taconite mine in the Penokee Range first spread, the issue became extremely polarized. Because the proposed mine lay in ceded territories, the Anishinaabe tribes in the basin claimed that the state had violated tribal sovereignty by failing to consult with them. Local residents responded with death threats against tribal members, and a swirl of local, state, and federal lawsuits, hearings, and threats marked the next four years. Debates over the mine became intense enough to swing a key election in Wisconsin. After the legislature had defeated a pro-mining bill by one vote in March 2012, pro-mining groups donated $15.6 million dollars in campaign contributions and lobbying fees to candidates who might support a mine.[32] The result was a change in control of the Wisconsin Senate (by one vote) after the November 2012 elections, giving the far right the power to rewrite Wisconsin laws.

In February 2013 a new mining bill exempting taconite mining from many of the state's water quality and wetlands standards was passed in Wisconsin.[33] The bill, written with the help of GTAC lobbyists, formally established the expansion of the iron mining industry as a policy of the state. This means that if there is a conflict between a provision of the iron mining laws and a provision in another state environmental law, the iron mining law overrides other state laws.

With the new law, the voice of the state became the only voice allowed in negotiations about permits. Local communities and the public lost the right to challenge state science and state permits. Contested case hearings—where the state had to face expert witnesses who might challenge their versions of the evidence—were outlawed. Citizen suits against a corporate or state employee alleged to be in violation of the metallic mining laws were also outlawed, even if the mine operator or state staff knowingly violated the law. In other words, the democratic processes by which outside voices could challenge state or corporate versions of scientific claims were outlawed, leaving in their place an echo chamber. Senator Fred Risser (D-Madison), the longest-serving state legislator in U.S. history, expressed the anger of many when he thundered in the senate, "This bill is the biggest giveaway of resources since the days of the railroad barons."[34]

The state of Wisconsin, however, cannot outlaw legal challenges from the Anishinaabe. The battle over the Mole Lake mine led to a U.S. Supreme Court decision in 2002 affirming the right of Indian nations to set and enforce their own clean air and water standards (working with the federal EPA). In other words, the state cannot set more relaxed standards. Even though a series of treaties between the federal government and sovereign Indian nations makes it clear formal consultation is required before environmental permits are issued, the Wisconsin governor and legislature decided to ignore those requirements when drafting the new law. In 2013, Scott Fitzgerald (R-Juneau), the senate majority leader, said that he had no plans of consulting with the Bad River tribe during the drafting process — a clear violation of tribal rights.[35]

When iron prices plummeted in early 2015 from a high of more than $180 per metric ton to under $60, GTAC backed out of the mine project, claiming that the Environmental Protection Agency was plotting to block the mine. The EPA disputed those concerns, pointing out that Susan Hedman, regional EPA administrator, had already denied the request of six tribes to evaluate the potential ecological risks of the mine. Yet rather than be silenced, the tribes partnered with wetlands ecologists to delineate the full extent of wetlands across the proposed mining site — and it was that ecological knowledge that eventually halted the mine, GTAC president Bill Williams eventually admitted. Williams was concerned that if his company was forced to protect wetlands, the mine would not be profitable. Even though the new state law seemed to allow destruction of wetlands in aid of mining, it was clear that the tribes would do their best to block any such action.[36]

Challenging Colonial Visions

In the late nineteenth and early twentieth centuries, North American capital and state interests brought mineral resources into production. From one perspective, that of the mining interests and industrial colonizers involved, it was a story of progress and human improvement: the advance of enterprising capitalists, civilization, and modernity into lands considered empty and unproductive. Urban governments long envisioned the north as a remote hinterland best suited for resource extraction. Yet the north is intimately connected to sites of industrial activity by animal migrations, by toxic mobilizations, and by historic legacies. Mining conflicts in the north continue to revolve around what kinds of relationships to natural and social communities will be supported by development—and who has the right to decide.

Mines in the Lake Superior region have long been a catalyst for Indigenous dislocation and dispossession. They remain what Timothy LeCain calls "a symbol of the political and economic power of outsiders to shape local environments in ways that serve the needs of national ambitions and global capital rather than those of local people."[37] Yet outsiders do not always win these conflicts. Like other northern regions, the Lake Superior basin was perceived by whites as an isolated and depopulated region, deeply in need of industrialization, progress, and tidy white communities. On the iron ranges of Lake Superior in the 1950s, corporate interests, national development priorities, and the international Cold War political economy converged to promote a single vision of the north. As we have seen in earlier chapters, many people were strongly exposed to that expansion, deeply concerned about fish, recreation, and forestry. Yet their concerns were overridden in the permitting process because planners agreed on what they believed to be a greater good: bringing industrialism to the remote north to fight the Cold War and to insure continued development.

Many people who favor new mines in the Lake Superior basin interpret the earlier iron mines as sources of economic prosperity. They find it easy to judge the environmental effects of proposed new mines in terms of what LeCain calls "scientifically-defined levels of human, and to a lesser degree, ecological toxicity. Supporters of the mine tend to put considerable faith in modern technology to limit the release of these chemicals and considerable trust in the involved companies to actually do so." But for the Anishinaabe, what constitutes clean water and healthy ecosystems emerges in relationships with wild rice, abundant fisheries, and water pure enough to drink. Their understanding of potential injuries from mining pollution is based on their cultural histories—and futures—with wild rice and healthy waters.

Conflicts over the meanings of mining's past continue to play an important role in current controversies over the sustainability of possible future mines. Decisions about sustainable mine planning are not purely scientific or technical decisions; at heart, they are social decisions. Mining companies now exploring the Lake Superior basin for new mines argue that while past mining practices may have caused some damage, the future will be different because the industry has embraced sustainability. New mines, they argue, will bear little resemblance to past mines. Furthermore, mining companies insist, mining is an essential part of sustainable development because it helps to fund economic development and environmental protection. A few sacrifice zones in remote regions will enable nations to experience the full benefits of neoliberal development.

Indigenous groups view these claims with skepticism. They point out that they have historically borne the brunt of sacrifices made in the name of economic development. Current mining industry discourse, they say, only replicates assumptions that places remote from urban centers are essentially barren and empty, rather than landscapes peopled with Indigenous communities.

In the Lake Superior basin, modern mining's technologies of what LeCain calls "mass destruction" produced a deeply scarred landscape. Those scars were cultural as well as ecological. Indigenous communities often bore the greatest burden from toxic wastes and social instability fostered by mining projects, but until recently, they rarely had much decision-making power in the planning process. Powerful tensions developed between metropolitan efforts to extract iron ore and Indigenous efforts to sustain wild rice, clean water, and abundant fisheries. Like the toxics mobilized by mining, those tensions have continuing legacies.

In 2011, the capture of the Wisconsin state government by the far right led to the evisceration of environmental protection. New policies attempted to silence public protests and marginalize challenges to a narrow range of technical expertise. The Indigenous peoples of the region, however, have federally protected rights to exert significant control over such developments—rights that cannot be eliminated by any state government. As a result, the Anishinaabe have become central to the attempts to sustain and protect these areas. Indigenous peoples of North America, and the globe, offer an alternative vision of the future—and on ceded territories, they have critical legal tools and strategies for protecting clean water and healthy watersheds.

Contested interpretations of the past continue to shape current conflicts. People who want the mines back point to a time when miners had good jobs, rarely mentioning the lung diseases that haunted the Iron Range, the bitter battles to win the few rights they had during the brief boom, the collapse of economies when the companies pulled out. Pro-mining individuals in Hurley told me in informal conversations that they feel that they can trust the mining companies, so there's no need for regulation or oversight. They remember a time when local impact funds created good schools, decent hospitals, well-maintained roads. But they forget that these benefits weren't just handed to them by the company. They were won, bitterly, in political fights led by unions that have since lost much of their power. Companies left to themselves never gave us anything, one resident of Minnesota's Iron Range said, concerned that new laws removed the protections that once marked the range.

Events across scales shape the most local processes within the basin. When the Asian building boom in 2011 forced global steel prices to new highs, what had been just a pile of useless rock to U.S. Steel became reframed as the nation's most important source of iron ore. Mining advocates insist the mine is inevitable in a global economy. "Only a primitive, backward people would stand in the way of our prosperity," one white woman from Hurley told me, complaining bitterly about the Bad River Band. But from the band's perspective, how can you destroy the water, the wild rice, the rivers, the slough, for a few jobs and a billionaire's profit? Water isn't a resource to be commodified; it's the blood at the heart of their place and life.

Advocates point out that forbidding taconite mines in the Lake Superior basin will only shift more mining to other places, where environmental and labor protections may be even weaker than in today's Wisconsin. And it's true that the proportion of world iron ore mined in Canada and the United States has dropped to only 3.5 percent in 2010. China, in contrast, mined 37.5 percent in 2010, and was still the world's largest importer. Yet who gets to decide which places, if any, are unsuitable for a mine? Who decides how to measure benefit versus harm? Are there going to be places where communities decide that the local, particular harms far outweigh the benefits on the scale of region or nation?

In a talk at Yale University in March 2012, western historian Richard White spoke of "incommensurate measures."[38] What's gained in resource development by one group cannot simply be measured against what's lost by another group, he warned. At one mining hearing, Richard White's incommensurate measures were in full force when pregnant women from the band spoke of their fears when they had to drink water poisoned by taconite mining.

Environmental history cannot tell us whether mining in a particular place should happen—that is a social decision, not a scientific or historical decision. But historical perspectives can remind us that there is nothing natural or inevitable about resource development. Resources are contingent and they change over time. Calling something a resource pulls it out of its intricate social and ecological relationships, isolates it in our gaze. Yet those isolations are illusions. We still live in intimate relationships with those elements, even if we think we don't. The language of inevitability masks the fact that government actions promote one vision of resources over another. So treaty rights and environmental quality must bend to the march of progress. What's hidden is the texture of the wild rice beds, the lake trout that swim through the waters of Lake Superior, the children of women poisoned by mercury, the asbestos released into the watershed by the processing of certain kinds of taconite deposits.

SEVEN

The Mysteries of Toxaphene
and Toxic Fish

THE PERSISTENT, BIOACCUMULATIVE chemical named toxaphene offers a case study on the history of toxic contamination in Lake Superior fish. How did chemicals such as toxaphene make their way into fish in the postwar era? How did governments and communities around the Great Lakes struggle to comprehend and then control these toxics? My goal is to explore the intersection of human culture with the pollutants that have made their way into water bodies—and the bodies of fish and the people who eat those fish—everywhere. Fish is a healthy source of protein that we're encouraged to eat, yet pregnant women face wrenching choices when they wonder whether the fish on their plate might harm their developing fetus. Fatty fish provide important nutritional benefits to the mother—but the fats could be contaminated with chemicals that could harm development. These aren't choices anyone should have to make.

In the late 1990s, Canadian scientists noticed that levels of toxaphene were rising in the fish of Lake Superior, even though the chemical had not been used near the lake for decades (see chapter 3).[1] Toxaphene had been banned in 1990, yet unlike other persistent bioaccumulative toxics, its levels were not dropping. Of all the Great Lakes, Lake Superior is easily the cleanest. So why would toxaphene be highest in this particular lake, in a region where the chemical had never been produced or used in agriculture?

What might toxaphene contamination mean for one of the great recovery stories of modern conservation: the restoration of Lake Superior fisheries? Even more pressing, what might that contamination mean for Ojibwe bands trying to restore culturally significant foods such as lake trout to their diets? When researchers examined toxic levels in fish and people, it became clear that people who ate fish from the Great Lakes were accumulating significant levels of banned toxic chemicals. During pregnancy, women passed these chemicals to their developing fetuses at their most vulnerable stages of development. The legacies of the past were becoming the body burdens of future generations. Because consumption of fish is an important vector of human exposure to endocrine-disrupting chemicals, toxics are particularly controversial in Indigenous communities in the United States and Canada where levels of fish consumption are high or where chemical industries have been sited close to Indigenous reserves and food sources.

Toxaphene refers to a group of turpentine-smelling chemicals made from pine oil and chlorine—two natural chemicals that are combined into a synthetic substance. Soon after being introduced into commerce after World War II, toxaphene was found to be toxic to fish, birds, and mammals. In the late 1940s, research was published showing its toxicity, and over the next two decades, studies indicated it to be mutagenic and carcinogenic in mammals.[2] After DDT was banned, toxaphene manufacturers began promoting toxaphene as a safe alternative. Research showing toxaphene's risks had been published for decades, yet because toxaphene was made from nature's own building blocks, customers believed it to be safer than DDT. Soon it was being mass produced and widely used as an insecticide, particularly in the cotton-growing industry in the American South.

Toxaphene became one of the most heavily used pesticides ever—as much as 46 million pounds a year during the height of its use in the 1970s, according to the U.S. Agency for Toxic Substances and Disease Registry. In the United States, most of that was sprayed on cotton and soybeans in the South, but it was registered for use in more than eight hundred products— for everything from tick control on livestock to the killing of unwanted fish species in lakes and ponds. Between 1964 and 1982, an estimated 500 million pounds of toxaphene was produced in the United States; production peaked in 1975 at 59.4 million pounds. As late as 2010, toxaphene was still available from eleven suppliers worldwide (including seven American suppliers). The total global production of toxaphene between 1950 and 1993 is unknown, but estimates range as high as 1.4 million tons.[3]

After research showed it to be mutagenic, teratogenic, and carcinogenic, the use of toxaphene was banned in the United States and Canada in
1990. But instead of destroying existing stocks of the chemical, manufacturers were allowed to continue production for overseas sales, and they soon
began marketing the chemical worldwide. Throughout Russia, China, and
Africa, the chemical found its widest use.[4]

Lake Trout Collapse

Meanwhile, back in North America, biologists, governments, and fishermen were busy trying to keep lake trout (*Salvelinus namaycush*) from extinction. Lake trout were once present in enormous populations within the
Great Lakes. Slow growing, they typically become sexually mature at seven
to ten years of age, and they can reach huge sizes in Lake Superior (figure 7.1). Voracious predators at the top of Lake Superior food chains, lake
trout had delighted fishermen for centuries. Even after other fish populations crashed under fishing pressure, pollution, and habitat loss, lake trout
had appeared to be surprisingly resilient. Some years their populations
dipped a bit, but they always seemed to recover.

In the mid-twentieth century, however, lake trout populations suddenly plummeted (figure 7.2). In 1944, the commercial catch of lake trout
in Wisconsin alone totaled more than six million pounds. A decade later,
only a few fish were caught, and by 1956 lake trout had vanished from most
of the Great Lakes. As they were top predators, the loss of lake trout had
rippling effects throughout aquatic food chains. Populations of nonnative
fish, such as alewives (*Alosa pseudoharengus*) and rainbow smelt (*Osmerus
mordax*), exploded when their predators vanished. In turn, as smelt and
alewife populations increased, they began to suppress plankton species
lower on the food chain, which in turn suppressed whitefish populations.[5]

Why did lake trout crash so suddenly? For decades, fisheries biologists have placed most of the blame on the sea lamprey (*Petromyzon marinus*), an invasive species from the Atlantic Ocean that, in the words of the
USGS, "quickly devastated the fish communities of the Great Lakes." Sea

Figure 7.1 Lake trout in Lake Superior can reach enormous sizes. Photograph from c. 1940–1965. (Wisconsin Historical Society, WHS-49575.)

lamprey attach to lake trout near their hearts and suck their bodily fluids (figure 7.3).

The historical narrative offered by fisheries biologists is that sea lamprey invaded the upper Great Lakes after modifications to the Welland Canal allowed marine organisms to make their way upstream past Niagara Falls. Once sea lamprey established themselves, they followed a pattern all too familiar from other exotic invasions: without predators, their populations skyrocketed, and they soon devastated their prey. But just in the nick of time, chemists and fisheries biologists managed to restore lake trout in

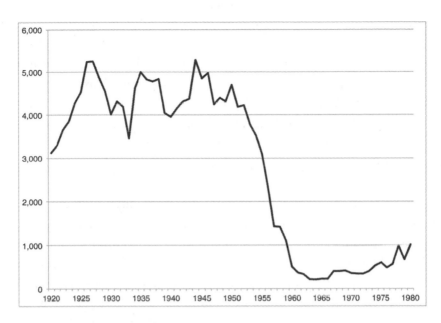

Figure 7.2 Commercial harvest of lake trout in Lake Superior, by thousands of pounds, 1920–1980. (Data from Global Great Lakes, University of Minnesota, http://www.globalgreatlakes.org.)

Figure 7.3 Sea lamprey attached to lake trout. (U.S. Geological Service image, via Wikimedia Commons.)

Lake Superior with the help of a synthetic chemical described below that kills developing lampreys without hurting too many young lake trout.[6]

Lake Superior lies at the top of a Great Lakes basin filled with examples of fisheries that had already collapsed in lakes that had become too polluted to support much aquatic life. Yet, as fisheries historian Margaret Beattie Bogue shows, the political chaos of different jurisdictions meant that few effective actions were taken to regulate the catch, protect spawning habitat, or clean up the nearshore environment.[7] On the land, the chaos of local, state, federal, and provincial laws and policies may have benefited forests, for it probably shaped an increased ecological diversity in the recovered forests. But in the water, that political fragmentation had very different effects, leading to a regulatory paralysis that thwarted effective action to prevent the collapse of the lake's fisheries.

Into this context swam the sea lamprey, an easy target for blame. But sea lamprey alone were never an entirely satisfactory explanation for the collapse of lake trout. First, the timing was off. Sea lamprey had been in Lake Ontario long before lake trout populations began to drop, and when the lamprey arrived in Lake Erie, they initially had little effect on lake trout. Similarly, commercial fishing pressures alone do not explain the collapse, because other fish, such as four-horned sculpin and burbot, also crashed at the same time, even though they were not being commercially fished. Finally, efforts to control sea lamprey and control harvests did not lead to recovery of breeding populations, except in Lake Superior. Hatcheries still stock all the lake trout that swim in the other Great Lakes.

Serious attempts to control sea lamprey began in 1950 with the installation of mechanical barriers that blocked most lamprey from getting up into the tributaries to spawn. But these barrier control measures were not perfect, and enough sea lamprey snuck through to continue hammering lake trout. In 1958, a chemical lampricide (and potent endocrine disruptor)—3-trifluoromethyl-4-nitrophenol (TFM)—was developed that killed larval lampreys in streams without killing adult trout. At the concentrations needed to kill lamprey, TFM did not seem to harm lake trout, so it seemed like a good choice. But the chemical did kill many stream invertebrates, which were essential for maintaining the health of fish populations.

In an attempt to control lamprey ammocoetes without devastating macro-invertebrates, fisheries biologists developed treatment protocols that called for tributary streams to be poisoned every three to five years, giving the invertebrates some time to recover before the lamprey recovered. Equally important, TFM prices were high because a single company (Bayer) controlled production, thus fisheries agencies had to innovate a variety of control options to save money. It meant, in effect, a form of what's now called "integrated pest management," which helps reduce the use of toxics and slow the evolution of resistance. There is no question but that chemical control was necessary for lake trout recovery. Yet chemical control alone was not sufficient. A combination of hatcheries, barriers, habitat restoration, toxic waste reductions, and fishing restrictions were important factors in the recovery.[8]

Lake trout eat other fish, putting them near the top of the food chain in Lake Superior and making them vulnerable to chemical bioaccumulation. Recent research has established a connection between dioxin levels, larvae mortality, and lake trout decline in the lower Great Lakes, particularly Lake Ontario. As described in chapter 3, dioxins are by-products of industrial processes; they typically form during the burning of chlorine-containing waste products or during herbicide production. Lake trout are extremely sensitive to early life stage mortality associated with dioxin exposure. At 30 parts per trillion, dioxin will begin to kill some lake trout larvae. At 100 parts per trillion, no lake trout larvae survive. Measurable levels of dioxins first showed up in Lake Ontario in the 1930s, and between 1950 and 1975 they were above 100 parts per trillion. This meant 100 percent mortality of larvae. Only hatchery fish could survive in the lake, and they did not survive for long. In Lake Superior, dioxin levels never reached the levels found in Lake Ontario, which may be part of the reason why breeding populations did manage to survive.[9]

Dioxins are not the only contaminants that affect lake trout. In the early 1980s, biologists discovered that Lake Superior lake trout were contaminated with high levels of the detritus of industrial civilization: PCBs, DDT and its metabolites, and dioxins, among others. Pollution had not been diluted into the deep lake but instead had become concentrated in the fish that people were eating. Grassroots fury at governments and corporations eventually led to a set of regulatory reforms that banned or strictly limited persistent organic pollutants, and levels of most fish tissue contaminants declined in the following decade. Those contamination levels, however, soon leveled off well above zero, with the potential for continuing effects on fish reproduction and health. Other contaminants were never banned, even though they clearly affected fish health. For example, phenols from pulp and paper mills continue to be widespread pollutants in lakes and rivers. Phenols affect the thyroid and reproductive hormones in fish, leading to problems with growth, sexual maturation, and immunity. But no studies have yet analyzed the possible effects these pollutants may have on entire fish populations.[10]

While the pollution in the lower Great Lakes—particularly Lakes Erie and Ontario—was obvious by the early decades of the twentieth century, most people thought Lake Superior remained pristine. Far from large industrial centers and surrounded by forests rather than chemical factories, Lake Superior had nevertheless become contaminated with a virtual soup of toxic chemicals that might affect fish reproduction and development—and the health of people who ate those fish.

How Toxaphene Got Around

By the late 1980s, breeding populations of lake trout appeared to have re-established themselves in Lake Superior, though not in the rest of the Great Lakes. So the finding that lake trout were contaminated with toxaphene unsettled biologists. Where was that toxaphene coming from? What effect might it have on fisheries in recovery?

Researchers initially suspected the culprit was the pulp mills lining the Canadian shores of Lake Superior near Thunder Bay, where deforestation of regional boreal forests had begun in the 1980s. The harvests supplied a growing paper industry, which dumped pulp mill wastes directly into Lake Superior. Those wastes contained chlorine and pine oils, which could combine under certain natural conditions to form toxaphene. But even in such inland lakes as Lake Siskiwit on Isle Royale—unaffected by toxins directly transported via water from Thunder Bay—contamination was high.[11] Researchers soon suspected that the chemical was coming not from local, contemporary sources, but instead from sources much more distant in time and place.

Evidence suggests that the chemical continues to be volatilized from old cotton fields in the American South, and that global wind currents may also be transporting toxaphene, which is still used in Africa, into Lake Superior and other boreal lakes—where it finds its way into fish and eventually into the people eating that fish. Once it falls into Lake Superior, it tends to concentrate, for the lake never gets warm enough to allow much toxaphene to go airborne again. Canadian researcher Terry Bidleman has documented toxaphene carried by the grasshopper effect as far north as the polar ice cap. Polar bears, seals, fish, and other animals store toxaphene in their fat.[12]

What risks do toxaphene and similar contaminants in Great Lakes fish pose for people who eat those fish? Among those most concerned are Indigenous communities who live along the shores of Lake Superior. Lake trout is a significant portion of their diets, and fishing is central to their cultural traditions. Many tribal health departments have urged members to eat more traditional or country food, wild food, to reduce the risks of diabetes, heart disease, and obesity from high-fructose corn syrup, white flour, and trans fats that make up the average American diet. But much traditional food, such as lake trout, is at the top of long food chains and therefore laden with synthetic chemicals that have bioaccumulated in the fat of prey species.[13]

Does eating that fish threaten the health of people? Health is more than just whether or not a person gets sick. Fish is particularly good for the health of fetuses and young children, and eating fish is of great cultural significance for the Anishinaabe. "When there are contaminants in subsistence foods, they pose much more than a physical health threat. It has an impact on spiritual, mental and emotional health as well," says Maxine Cole, coordinator of the EAGLE Project, an environmental health-research undertaking.[14] The potential contamination of fish forces communities to make trade-offs between their beliefs and possible harm to themselves and their future children. How much fish do you eat when it's culturally important? How much do you eat when you're pregnant?

The Lac Courte Oreilles Band in northern Wisconsin, with about six thousand enrolled members, harvests between 1,900 and 2,500 fish during spring spearfishing season. To decide which lakes are safe for spearfishing, tribal managers and individual anglers turn to risk-exposure maps created by the Great Lakes Indian Fish and Wildlife Commission. These maps reveal which lakes have the highest levels of methylmercury in fish tissue that year. At the same time, tribal members are encouraged to catch and eat mostly smaller fish, both during the spearfishing season and throughout the year. Because families freeze so many fish to eat later, they are also taught to label each bag with the weight and species of fish, along with where it was caught, to help them monitor their families' exposure to methylmercury and other contaminants."[15]

In Canada and the United States (as in Europe as well), governments address the potential risks posed by toxic contamination in fish by issuing advisories — lists of fish that people should not eat. Currently, the Canadian government advises citizens in the Lake Superior watershed to avoid eating lake trout altogether because of toxaphene contamination. Across the border in the United States, levels are equally high, but for complex political reasons, no advisories warn against consumption of fish contaminated with toxaphene. The fish ignore national borders: within moments, a fish too toxic to eat becomes perfectly safe when it swims a few yards across an invisible line. Advisories assume a "normal" amount of fish consumption, but the definition of "normal" is problematic, for it has been based on an imagined white, middle-class adult male. Advisories have ignored much higher levels of fish-eating in Indigenous societies. Advisories also assume all bodies respond to a given level of contaminant in exactly the same way. But women respond differently from men, children respond differently from adults, and fetuses respond most intensely of all. Finally, advisories assume alternatives exist, so people can just drive to the store and buy something else. This ignores the lived reality of many subsistence communities.[16]

Changing Understandings of Risk

As understandings of bioaccumulation and biomagnification emerged, how did scientists change their understandings of the effects of contaminants on people? Before World War II toxicologists developed a set of frameworks for understanding the potential effects of chemical contamination on human physiology. Toxicologists believed that most potential poisons had a threshold value below which the substance was unlikely to affect an individual's functioning or physiology. Similar to the "dilution is the solution to pollution" approach, toxicology assumed "the dose makes the poison." Human bodies, toxicologists believed, could accommodate some degree of toxic exposure as long as the exposure was below a certain threshold. Although this dose-makes-the-poison concept is true for many contaminants, it is not true for others, particularly for hormone disruptors. At low concentrations, hormones normally stimulate receptors and set in motion a series of physiological responses (such as breast development at puberty). At high concentrations, however, hormones can saturate receptors, thus inhibiting their pathways. Low doses of hormone disruptors such as DDT or DES might harm individuals even though higher doses might not.[17]

But how do we know potential risks, given that people are exposed to many different sources of contamination? The most striking thing about the human health effects of eating Great Lakes fish is how confused and contradictory the results have become, even after decades of research. Thousands of studies have been conducted over the last two decades, and the results of those studies are anything but clear. There is no doubt that high levels of contaminants in fish—for example, mercury, toxaphene, or PCBs—increase the risk of birth defects, miscarriage, and neurological damage. But what about lower levels? Here the results are contradictory, in part because the Great Lakes are awash in a bath of different toxic chemicals, and these chemicals often show opposing effects. At low levels, they may cancel each other out or they may synergize and heighten effects.

Take the case of pregnant women and fish consumption. The brains of developing fetuses benefit from omega-3 fatty acids, and the typical modern diet is deficient in omega-3s, which are richest in fatty, cold-water fish.

Eating fatty fish is an excellent way to make omega-3 fatty acids available for the developing fetus. But it is precisely these fats in fish that also carry the lipophilic toxins, putting a pregnant woman—or a woman who might someday hope to get pregnant—into a bind. Does she eat those fatty fish from Lake Superior, secure in the knowledge that their omega-3 fatty acids are helping neurological connections form? Or does she shun them? What does a woman do? And what does a population do? Is the health advice that makes the most sense for an individual necessarily the best policy advice for an entire population that has to balance multiple risks and multiple costs?

The EPA still assures folks that "the average person in the Great Lakes basin may not be at risk of experiencing adverse health effects from exposure to contaminants through the consumption of fish."[18] It is assuming a very low average level of fish consumption—a few ounces a week. But as the EPA admits, some people eat a lot more fish than that, and they eat fish high on the food chain, often from highly contaminated waters. And, most important, all people don't face the same risk from the same plate of fish: the developing fetus and young children are at greater risk than adults.

Fish Advisories

Fish advisories have become so central to modern pollution regulation that they are essentially naturalized. They seem purely rational, scientific, and authoritative to regulators. A great deal of debate in the scientific and policy literature swirls around fish advisories, but always at the edges: should the tolerances be 0.1 or 0.2 parts per million per day? Should they be calculated based on an average fish consumption of 0.23 ounces of fish per day or half an ounce per day? These are interesting scientific debates, but they ignore important questions: How did this advisory approach develop, and how did it become naturalized as the most rational approach to fish contamination? What were the assumptions that guided its evolution and acceptance, and who has challenged these assumptions?

The industrialization of urban centers in the nineteenth century brought the problem of waterborne pollution to reformers' attention. Yet fecal wastes from people and livestock had long been a concern for

city dwellers. Greek and Roman, Islamic, and Indus River societies had
invented elaborate water sanitation systems that did far more to protect
public health than the systems that were pieced precariously together in
northern European cities around the early nineteenth century. There was
a greatly increased risk of exposure to waterborne bacterial and viral diseases as well as industrial pollutants. The London cholera epidemics of
1848–1849 and 1853–1854 resulted in parliamentary reports examining
sanitation and water quality, and a Board of Health report of 1850 called
for a publicly administered water supply for London. Victorian reformers
pressed for municipal waterworks, which helped to ensure cleaner drinking water for urban residents. Sewage treatment came much later.[19]

While these reforms removed fecal matter from drinking water, they
failed to limit industrial pollutants dumped into urban waters. The historian Leslie Tomory describes the pollution in 1800s London from the growing manufactured gas industry. Fishermen in the Thames were particularly
furious, for the toxic wastes destroyed their ability to harvest and sell fish.
Pollutants destroyed fish habitat and made the few fish that survived taste
too nasty for human consumption. Fishermen took the industries to court
but lost their cases—not because they could not prove the gas industry was
polluting the water and poisoning the fish for human consumption, but
because the courts refused to restrict a growing gas industry that seemed to
promise so many benefits to the city. What did a few poisoned fish matter
when the lights of London were at stake?[20] Fish advisories did not emerge
perhaps because fish were so disgusting to the eye and nose that consumers
were at little risk of eating toxic chemicals.

Similarly, in Canada and the United States in the late nineteenth
century, paper companies dumped stinking pulp wastes into the spawning beds of Lake Superior fish. Fishermen and consumers complained,
but governments refused to restrict the dumping of waste, preferring instead to recommend chlorination (which protected humans from bacterial
contamination but only increased the harm to fish stocks). The fish that
survived smelled revolting, and consumers refused to eat them. (Ironically, this helped protect some of the fish stocks, which were released from
the heavy commercial fishing pressures that devastated fish populations

in rivers without pulp mills.)[21] In the Great Lakes region, just as in the Thames River watershed, two industries were in conflict, and courts and regulators reluctant to limit industrial expansion sided with what they saw as the more modern, profitable industry. Because it seemed clear to the eye and nose which fish were polluted, governments assumed that consumers would protect themselves from toxicity by avoiding the fish that stank.

This assumption was challenged in 1920s New York, when sewer lines polluted oyster beds but the contamination was not obvious to consumers. After an outbreak of cholera was traced to oyster consumption, a lively debate in the public health journals arose about the risks of consuming seafood when invisible toxics lurked in the water. New York City commissioners decided to close the oyster beds rather than attempting to stem the source of pollutants. The hygiene boards also recommended surveillance: oyster farmers should test their own oysters, publishing the results so consumers would regain confidence about the purity of seafood.[22]

These early episodes reveal a growing awareness of pollution making its way from industrial centers into the broader environment, and from there back into people. Governments developed an approach to pollution that relied on surveillance to monitor contamination and isolate people from the sources of that contamination. This approach assumed that planners could separate people from the sources of contamination that fish encountered, and protect people from the waste products of industrialization by having people avoid contaminated food.

This approach was challenged in the 1960s and 1970s, when evidence emerged that toxic chemicals could poison people distant in both time and space from the sources of contamination. Perhaps the most infamous example is mercury poisoning in Minamata, Japan. From 1932 to 1968, Chisso Corporation—one of Japan's most profitable and powerful corporations—released industrial wastewater from a chemical factory in Minamata City.[23] Methylmercury was a by-product of the manufacturing process that was released with the wastewater into the bay.

In April 1956 a five-year-old girl entered the factory hospital with severe brain damage, unable to walk or speak coherently. A few days later, her sister entered the hospital with the same symptoms. Soon hundreds of their neighbors were affected as well. In May, the city government and various medical practitioners formed the "Strange Disease Countermeasures Committee." The committee uncovered "surprising anecdotal evidence of the strange behaviour of cats and other wildlife in the areas surrounding patients' homes." Cats had convulsions then seemed to go mad and die. Locals called it the "cat dancing disease." Historian Brett Walker quotes one Minamata resident who noticed that local cats—which had eaten the fish from the bay—went mad first: "Then cats started to go mad. We couldn't believe our eyes when we saw them running around and around. . . . In the end they jumped into the sea and drowned. All the cats in the fishing villages around Kyakken died in this way. Then people began to show the same symptoms. The poisoned fish did their job damn well."[24] The committee asked researchers from Kumamoto University Medical School to investigate, and in October 1956 they issued a public report that implicated heavy metal poisoning caused by eating contaminated fish from the bay.

Chisso refused to admit that methylmercury was in its wastewater, and it refused to allow samples to be tested. In 1958, the company diverted waste water from the bay to the Minamata River, exposing more residents. Dr. Hajim Hosokawa, a doctor employed by the company, began feeding cats the same effluent that his employer was dumping into the water, and his cats fell ill with similar symptoms. When he took his results to Chisso's management, the company refused to allow him to continue his research and suppressed his results. Not until 1968 did the government declare that the disease cluster was caused by pollution from the Chisso factory.[25] Years passed before the source of pollution was finally regulated.

Minamata revealed that toxic contamination could have effects that were difficult to trace because they were distant in time. The chemicals poisoned not only the people who ate the fish but also a generation removed. Pregnant women who ate those fish had given birth to children who bore the effects. Surveillance as an approach to pollution relied on transparency, yet industries had few incentives to reveal the contents of their toxic wastes or research studies and many incentives to hide the results.

Experiences with PCBs suggested that contamination might be distant not just in time but also distant in space from sources of contamination.

In 1964 the Swedish chemist Sören Jensen was analyzing DDT levels in human blood when certain chemical compounds kept recurring in his samples, interfering with his analyses. After determining that the chemical was synthetic, he eventually established that it was chlorine-based (like toxaphene) and similar to DDT. Yet he realized it could not be a chlorinated pesticide, because museum specimens allowed him to determine that it appeared in white-tailed eagles (fish-eating birds) by 1942, several years before chlorinated pesticides were widely used.[26] After sampling two hundred fish from around Sweden, Jensen realized that all of the country's water and its adjacent coasts were contaminated. "Even hair samples taken from his wife and three children showed traces of the compound, with the highest levels in his nursing infant daughter. The mystery pollutant was everywhere he looked." By 1966 he realized that he was dealing with the industrial coolant and lubricant known as polychlorinated biphenyls, or PCBs. PCBs accumulate in water and soil and could be transported globally through atmospheric transport and through the migration of birds and fish. "The circle was closed," Jensen said. In 1972, Jensen noted that "since the discovery of PCB accumulation in nature was published in 1966, the presence of PCB has been reported in organisms from all over the world."[27] In 1972 Sweden essentially banned PCBs, and other nations followed suit, with Canada banning their production in 1978 and the United States in 1979.

After Jensen published his results in 1966, American and Canadian researchers examined Great Lakes fish and realized that they were also contaminated with PCBs. In response, the first Great Lakes fish advisories were issued in 1971. Over the next decade, jurisdictions within the Great Lakes watershed issued their own advisories, developing different policies and advisories based on different chemicals and different quantitative approaches. A fish might be safe in Michigan, but if it swam over the border into Wisconsin waters, it became too toxic for human consumption.

As of 2011, all fifty states, the District of Columbia, and five Native American tribes had a total of 5,627 fish advisories in place for 4,821 water bodies. This means that before eating a mouthful of fish, each consumer was supposed to check all these fish advisories on all these water bodies, to make sure she wasn't poisoning herself. In 2010, fish from only 2 percent of U.S. river miles were listed as safe. The other 98 percent of river miles? Those fish might be toxic. Or they might not be, and the burden of determining potential risks versus benefits was on each consumer.[28]

Critiques of fish advisories are becoming common in the social science literature. Such critiques typically focus on the poor communication of risks to target audiences, calling for more culturally appropriate advisories.[29] Yet, as several geographers have pointed out, fish advisories reproduce problematic neoliberal assumptions about environmental policy. Environmental risks that stem from industrial production are regulated by telling individuals to change their personal consumption rather than by telling industry to undertake systemic reforms.[30].

Fish advisories rely on quantitative risk assessments, which assume that risk is an unavoidable fact of modern life, something to be managed rather than eliminated. Risk assessors estimate the size of a given risk from a given chemical, and risk managers decide whether that risk is acceptable. This process relies on estimates of how an average person can be exposed to a particular toxic chemical without suffering significant harm. Harm is typically defined as getting cancer, although other endpoints such as reproductive failure are possible to assess. Risk assessors then manage pollution by permitting chemicals to be used and released, just so discharges do not exceed a standard of contamination deemed to be an acceptable trade-off for economic gains.[31]

While fish advisories may seem like a purely rational response to contamination trade-offs, the geographer Becky Mansfield suggests that they are "a form of gendered biopolitics of responsibility for population security." Because advisories "combine elements of reproductive politics, including the medicalization of pregnancy, the production of fetal personhood, and enduring notions about 'good' and 'bad' mothers" they represent "a gendered technology of biopolitics that intensifies the self-disciplining of women as mothers of potential, future children."[32]

Not just current mothers are targeted by these advisories; all women who might someday become pregnant are also urged to "self-discipline" themselves to protect future populations. For example, the European Union states that young girls must follow its advice regarding dioxins because of the effects of dioxin on the developing male reproductive system. Girls, in other words, must discipline themselves "to protect their future unborn male children. Seafood consumption advisories discipline women as mothers for more than half of their lives—whether or not they have children and during their own girlhood. Seafood consumption advisories are not just an example of the self-disciplining demands of intensive motherhood but an important extension of these demands throughout the life course and in the name of population security." Mansfield argues that "laying blame for the health problems of environmental pollution at the feet of 'bad mothers' is less a rational solution than it is an intensification of the neoliberal biopolitics of reproduction and mothering."[33]

Fish advisories conceptualize the problem of contamination as possible harm to individual human physical health. But this approach, the legal scholar Catherine O'Neill argues, "fails to appreciate the cultural dimension of the harm," for it ignores the "integral role of fish, fishing, and fish consumption in the lives" of many Indigenous peoples. Quantitative risk assessment attempts to use quantification to escape from cultural conflicts, but as O'Neill points out, "Far from permitting humans to 'get beyond a clash of sacred values,' quantitative risk assessment and related analytic approaches simply instate one view of the sacred."[34]

Risk assessments start by calculating a level of exposure based on estimates of average or normal fish consumption. For example, until recently the U.S. Environmental Protection Agency used in its risk assessment calculations the figure of 0.23 ounces per day of fish. This figure came from a single diet recall study in the mid-1970s, which calculated that the average fish-eating American ate half an ounce of fish per day. To estimate exposures to fish, the EPA recalculated that estimate to exclude marine species and to include people who eat no fish at all—thus coming up with 0.23 ounces per day (which comes to less than one serving of fish per month). This figure is absurdly low for tribal members from the Great Lakes region or the Pacific Northwest, who routinely eat more than 150 times the risk assessment estimate of fish. For example, the Michigan Great Lakes tribes of the Bay Mills, Grand Traverse, and Sault Ste. Marie bands all have a long

and carefully documented fishing culture. Tribal records show that historically, tribal members ate over two pounds per day of fish. Interviews done
in the 1990s found that 1 ounce per day over the full year was the average for
tribal members who lived off-reservation; on reservation Indians ate even
more. Yet when the 0.23 ounces per day figure was challenged in court by
tribal attorneys, the judge dismissed the tribal evidence as "anecdotal" and
"historical."[35]

Why do these numbers matter? The greater the estimates of human
fish consumption, the tighter the standard becomes for regulating point
discharge of toxic chemicals into surface waters. When agencies assume
that people eat little fish, they allow more toxic emissions. This creates a
"suppression effect." People eat less fish because fish advisories warn them
to do so—but then those lowered consumption levels are used as an excuse
to allow even more contamination. The National Environmental Justice
Advisory Council in its 2002 report on fish advisories argued that "when
agencies set environmental standards using a fish consumption rate based
upon an artificially diminished consumption level, they may set in motion a
downward spiral whereby the resulting standards permit further contamination and/or depletion of the fish and aquatic resources."[36]

The past also influences Indigenous community responses to agency
fish advisories. The anthropologist Valoree Gagnon found in her work with
the Keweenaw Bay Indian Community on Lake Superior that many tribal
members were suspicious of fish advisories issued by the state of Michigan.
The tribe had taken the state to court in the 1970s over its treaty right to
fish, and it had won the case in the Michigan Supreme Court decision of
1971 (*People v. Jondreau*). Shortly afterwards, the state issued fish advisories that told citizens not to eat fish caught by tribal members because it was
contaminated. The tribe's commercial fishing industry collapsed. Tribal
members suspected that the state had retaliated against them, but they also
worried about the contamination in the fish and what eating that fish—
now the only affordable source of protein for the community—might do to
their children.[37]

Risk assessment, in essence, has meant that society hands over regulatory decision making to a technological elite, whose members alone can

understand the increasingly arcane basis and data analysis of quantitative risk assessment. The quantifications appear precise because they are so complex and difficult to understand, yet assumptions within these models are typically unquestioned by the public. Communities suspicious of risk assessments have usually been dismissed as ignorant and largely excluded from the decision-making process. Quantitative risk assessment assumes that a certain amount of risk is acceptable. Yet with chemical releases, the risks are rarely distributed equally or equitably among the population. Pregnant women—or any female who might someday become pregnant—and Indigenous communities have often borne a substantial share of the risks from toxic releases.

Environmental justice advocates do not argue that agencies should simply stop issuing fish advisories. As the writer Paul Rauber notes, "Giving up local fish would mean giving up her culture, and while the effects of eating tainted fish are chronic, subtle, and often hard to separate from the manifold ailments of poverty, the effects of losing one's culture are there for anyone to see: alcoholism, broken families, drifting children. . . . In the end, the only problem fish advisories solve is that of informed consent. People have a right to know the risks involved in what they eat, but they have an even more fundamental right not to face those risks in the first place."[38]

Indigenous justice advocates do argue, however, that agencies should stop relying on fish advisories as a solution to contamination from fish consumption. The Environmental Justice Advisory Council report argued: "Finally, the health of humans and the health of aquatic ecosystems are intimately related, such that compromised aquatic ecosystems are of concern in and of themselves, with the contamination of fish, aquatic plants, and wildlife but some of the devastating effects. Water of sufficient quality and quantity is vital to sustain all life. To allow waters to be degraded and depleted is to undermine health, traditions, cultures, and economies. To allow waters to be degraded and depleted is to neglect obligations, including the obligation to sustain tribal homelands as contemplated by federal Indian treaties and other laws."[39]

Writer Madeline Fischer asks, "How did we reach this place where one of our healthiest foods has grown so complicated?"[40] A century of industrialization and environmental pollution is part of the answer. We eat fish today, yet the fish bear the traces of environmental history, contaminated with chemicals that were banned long ago. Fish consumption advisories are useful bits of information, but they fail to address the fundamental inequalities that make some people more susceptible to contamination than others. While the processes that shape watershed change have global roots, the effects are profoundly local. The toxaphene from Chinese, Russian, and African fields accumulates in the fish that swim under my cliff. Mountain-top removal mining for China's iron ore demands would release toxics, such as mercury, into the bodies of local fish, and from there they would make their way into local children. Global processes—financial markets, building booms, industrial farming practices in places with few environmental regulations—become local in the most intimate ways, as toxics from those distant industrial processes accumulate within our bodies.

EIGHT

The Great Lakes Water Quality Agreements

IN JUNE 1972 MY MOTHER PILED her five children, some friends, and our dog into the van to drive across the country, camping all the way. Hurricane Agnes nipped at our heels, flooding our Maryland suburb just as we drove away. High winds collapsed the tents at our Ohio campground on the shores of Lake Erie, where the lake was green with algae as far as we could see. When we reached my grandparents' house in the Chicago suburbs, we hopped out of the car and ran down to the community beach along Lake Michigan. There we were confronted with miles of dead alewives, alien fish rotting on the shores of our inland sea. The stink was overwhelming. Flies rose from the carcasses, silvery, peeling bits of death swinging back and forth on the waves, smacking into our sneakered toes. It reminded me of the corpses in *Night of the Living Dead,* a movie that kept me up at night for years. The alewife corpses seemed like the cranky undead, rising from the scum of the once Great Lakes (figure 8.1).

Four decades later, I now live on the shores of Lake Superior. Instead of rotting piles of invasive fish, local anglers reel in native lake trout. Bald eagles nest in the white pines above us. Kingfishers rattle when we launch our kayaks at the Lost Creek state natural area, where otters dip and dive and merlin nest. Once-threatened herring gulls expand their nesting colony along the cliffs under my cabin. When we kayak by in the evening, we swing out away from the cliffs so we don't disturb the gulls on their nests or distract the eagles who swoop in to feed their young.

On the surface, this is a miracle of conservation recovery. Dead fish and green lakes were common sights before the Clean Water Act and the Great Lakes Water Quality Agreement were signed. As the environmental historian William Ashworth has noted, many people had given up on the Great Lakes by 1969 when the Cuyahoga River caught on fire. David

Figure 8.1 Cleaning dead alewife off Chicago shoreline following great die-off of June 1967. (Image ID: fizh1403, NOAA's Historic Fisheries Collection.)

Schindler, one of Canada's most influential environmental scientists, writes that the international community finally took action after scientists discovered high levels of persistent contaminants in Great Lakes fish in the 1970s: "In brief, science and policy both responded to the harmful effects of contaminants to keep species from becoming extinct and humans and other top carnivores from being extensively harmed. I believe that this is one of environmental science's success stories."[1] And indeed it is—but legacy contaminants persist and new ones threaten these recoveries.

Native lake trout and the wildlife that eat them have indeed returned to Lake Superior. But while their populations have recovered from the crashes of the mid-twentieth century, as individuals they're still suffering deformities, reproductive failure, and endocrine disruption—as are people who eat the tainted fish in the Great Lakes. Levels of DDT, PCBs and other banned chemicals dropped steeply after they were removed from commerce, but they stabilized at amounts high enough to cause reproductive harm. Even more troubling, levels of emerging contaminants, such as flame retardants, are rising in Great Lakes food chains, including in the breast milk of human mothers at the top of those food chains.

Sediments contaminated with legacy pollutants still lurk in estuaries and river mouths, and storms churn those old contaminants back into the water column (figure 8.2). The past refuses to stay in the past. Winds and rains bring atmospheric contaminants, mercury among them, released from distant places into Lake Superior, and we bear the harms of distant decisions within our bodies. Fish are excellent sentinels of the effects of contaminants on global ecosystems—and of our often fragmented responses to those contaminants.

Concerns That Crossed National Borders

By the early 1970s, filthy conditions in the Great Lakes stimulated grassroots concern in both Canada and the United States. When Congress passed the amendments to the Federal Water Pollution Control Act (known as the Clean Water Act) in 1972, overriding Richard Nixon's veto, many of the nation's waterways had become "foul, scum-covered sumps for fac-

Figure 8.2 Contaminated sediment warning along the waterfront in Ashland, Wisconsin. (Kate Golden/Wisconsin Center for Investigative Journalism.)

tory discharges and untreated sewage."[2] The Clean Water Act radically expanded the federal role in cleaning up existing pollution and preventing point sources of pollution—pollution from a single identifiable source including factories. Sharp drops in pollution from sewage and industry resulted. In Canada, the Water Act of 1970 similarly expanded the federal role in consulting on provincial oversight of water quality. But because the Great Lakes were shared waters, the two nations had to coordinate their efforts, according to the terms of the 1909 Boundary Waters Treaty.

The Boundary Waters Treaty had come about because Canadians and Americans shared watery boundaries. By the late nineteenth century, the two nations were arguing over a host of pollution and water quality cases on their shared waters, including industrial development on the Lake of the Woods, hydropower developments on the Niagara River and on the St. Marys River coming out of Lake Superior, and irrigation withdrawals between Alberta and Montana. Canadians were quite eager to control pollution from Americans when it crossed into their nation but less eager to control their own industries. The same was true for Americans. In 1905, the countries agreed to establish the International Waterways Commission with representatives from both nations. Four years later the United States and Britain signed the Boundary Waters Treaty. The treaty created the International Joint Commission to study and regulate transboundary waters.[3] Both nations recognized that regional governments were doing little to protect water resources when it meant restricting economic development. But rather than usurp state or provincial control through federal-level regulation, the International Joint Commission turned to scientific surveillance and research.

By the 1960s, the failures of research and cooperative pragmatism to control Great Lakes pollution were becoming painfully evident. As conditions worsened, the governments of Canada and the United States requested that the IJC report on whether the shared waters were "being polluted on either side of the boundary to the injury of health and property on the other, contrary to the 1909 Boundary Waters Treaty." If so, the commission was to identify the causes and recommend remedial or other measures to address the problem. In 1970, the IJC reported that pollution was indeed occurring "on both sides of the boundary to the injury of health and property on the other side."[4]

In 1972 Canada and the United States signed the Great Lakes Water Quality Agreement, a revolutionary agreement that has become the "cornerstone of U.S.-Canadian cooperative efforts on Great Lakes water quality issues."[5] Intended to coordinate the actions of Canada and the United States in controlling pollution, the Great Lakes Water Quality Agreement's purpose is to restore and maintain the "chemical, physical, and biological integrity" of Great Lakes waters. Within the United States, the Environmental Protection Agency (EPA) is charged with coordinating activities to fulfill the agreement; in Canada, Environment Canada plays the same role. The agreement was groundbreaking in its focus on cleaning up existing pollution and preventing new pollutants, but the IJC has no authority to force the two nations to implement recommendations. Therefore, when Canada or the United States refuses to abide by the Great Lakes

Water Quality Agreement (in its various revisions), very little happens in response—besides calls for more research.

The Great Lakes Water Quality Agreement (hereafter the Agreement) signed in 1972 focused largely on phosphorus pollutants, because hyper-eutrophication from excess nutrient loading was a clear and compelling problem. And indeed, in its initial years, under IJC pressure, phosphorus in sewage effluent was controlled, and the lakes stopped turning green every summer from algae blooms. But the IJC was well aware that limiting phosphorus from sewage alone would not address even the eutrophication problem, much less the larger issues of toxics. Phosphorus from industrial products—particularly detergents—was much harder to control because industry vigorously resisted restrictions that might affect profits. The IJC had little power to force compliance, thus the governments relied instead on voluntary agreements, which were partial and slow.

In its annual reports on the Great Lakes Water Quality Agreements, the IJC commissioners made their frustrations clear, even though the tone was initially constrained. For example, in the fifth report (1977), the commissioners noted that "progress towards the goals of the Agreement continues to be slow and uneven." The commissioners expressed concern "with the increasing evidence of the dangers of toxic chemicals in the lakes and with the failure to implement enforcement regulations on many municipal and industrial sources of pollution."[6] That same year, frustrated that the nations would not meet toxics reduction goals set in 1972 by the Agreement, the two nations agreed to negotiate a revised agreement. Love Canal—another Great Lakes toxic tragedy—had shot into national consciousness in the summer of 1978, and toxics were no longer a topic for scientists alone to worry about.

In response to the furor raised by Love Canal, the 1978 Agreement shifted focus from conventional pollutants to persistent toxic substances, such as DDT, PCBs, mercury, and radioactive waste. Indeed, the 1978 Agreement called persistent bioaccumulative toxics the "most urgent problem" facing the lakes. The United States and Canada committed not just to *reducing* persistent toxic pollutants but to achieving their virtual *elimination* throughout the Great Lakes ecosystem. The Agreement went so far as to ban the discharge of thirty-two persistent toxic substances, including PCBs, inorganic metals, and DDT, stating quite clearly that "toxic substances in toxic amounts shall be prohibited and the discharge of any or all persistent toxic substances shall be virtually eliminated."[7]

Equally significant, the 1978 Agreement included what the commissioners called an ecosystem approach. The ecosystem, according to the Agreement, was the interacting components of air, land, water, and living organisms, including humans, within the drainage basin of the Great Lakes and the international section of the St. Lawrence River.[8] Restoring water quality and eliminating pollutants meant focusing not just on single sources but on the entire ecosystem.

In July 1982, the IJC released its first biennial report under the 1978 Agreement. Instead of a glowing celebration of the ecosystem approach and the efforts of both governments, the commission offered what one scholar called "the most biting critique of governmental performance which the IJC had issued in its seventy year history." Canadian studies scholar Joseph Jockel notes that language in the report came "as close as diplomatic language can permit to accusing the United States Government (in particular its Chief Executive) of hypocrisy and failure to fulfill its international obligations."[9]

In its report the IJC argued that "bold and innovative" approaches were needed to deal with challenges posed by complex chemical mixtures with cumulative effects. The IJC acknowledged that the two nations that were party to the Agreement (known in the reports as the Parties) had made some progress in decreasing DDT, mercury, and PCBs. But the commissioners complained that "the response to control measures is not, as yet, very dramatic" and certain banned chemicals such as dieldrin were not yet decreasing in the lakes.[10]

In particular, the IJC was frustrated that the Parties had failed to meet the most basic elements of the 1978 Agreement: drawing up lists of pollutants entering the Great Lakes in order to set timetables for their elimination. Because industry had insisted on the need to protect trade secrets, companies had refused to reveal the content of their chemical releases into the Great Lakes. A stumbling block quickly became lack of knowledge: there was no way for the Parties to determine "the chemicals manufactured, transported, or discharged into the basin."[11] Without knowing what was being released into the Great Lakes, how could it be cleaned up? The EPA was infuriated by the IJC's critique and ready to scrap the entire process. In turn, the IJC was angered by the EPA's unwillingness to set concrete timetables for cleanup and establish accountability rather than put forth vague assurances of good intentions. Over the next two decades, these basic issues remained fault lines: the IJC wanted timetables and concrete plans; the Parties refused because strict timetables might imperil their voluntary agreements with industry.

In its 1984 biennial report the IJC continued to criticize the Parties, reiterating once again that "there has been limited success in coming to grips with the overall problem of toxics in the Great Lakes basin." Even as the two nations that were party to the agreement endlessly discussed voluntary plans for removing current chemicals, new chemicals appeared in complex mixtures, with unknown effects on water quality and complex potential for injury to wildlife and human health. The report read: "New chemicals are being introduced continually. Even if only a few are known to be harmful, it is becoming increasingly apparent that their individual, combined and long-term effects do present serious environmental problems." Yet the nations refused to create a process that evaluated new chemicals *before* their release into the Great Lakes, an approach the IJC insisted was necessary to prevent future, expensive cleanup projects.[12]

The IJC was frustrated, as well, that the two nations continued to rely on standard toxicological testing based on assimilative capacity models. The IJC pointed out that while assimilative capacity might make sense for bacteria deposited in healthy, flowing waters where replenished oxygen could help break down dangerous bacteria, such models no longer worked for the new synthetic contaminants, particularly the persistent, bioaccumulative toxics. Even very low levels that were undetectable in standard water samples could still "exert adverse biological effects at dilute concentrations and bioaccumulate in aquatic organisms in the food chain to toxic proportions. These effects may include impaired reproductive processes and birth defects, neurological dysfunctions and behavioral aberrations, abnormal growth patterns including tumours and neoplasm production and reduced immunity capacities."[13]

Hormone effects of new contaminants disturbed the IJC in 1984—well before endocrine disruptors were on the radar screen of most environmentalists. The EPA and Environment Canada were essentially pollution-permitting agencies, not pollution-banning agencies. Both agencies had the task of issuing permits to allow companies to continue polluting—but only beneath a certain threshold that could be broken down in the environment. Government scientists created complete assimilative capacity models to calculate "safe" levels of pollutants so they could then issue permits for releasing those pollutants. But with persistent pollutants "there are limits to what technical and scientific programs [of control] can accomplish," the IJC warned. Bans would be necessary, not just permits allowing industries to continue to pollute.[14] It was a revolutionary perspective, essentially identical to the warnings of endocrine disruptor scientists three decades later. Concerns about low dose effects, fetal effects, sentinel warnings (early warning systems of reproductive harm): these were all present in the IJC report of 1984.

And clearly, the ideas presented in the report were too radical for the nations, which had enough trouble calculating thresholds for pollution permits. Ten years later, in 1994, the IJC was repeating these same arguments: relying on assimilative capacity models would simply not work for persistent bioaccumulative toxics that had the potential to disrupt hormones. Such models assumed that the Parties could allow industries to release pollutants based on "ambient pollution levels below which residual risks are considered minimal." The IJC repeated, in biennial report after biennial report: pollution approaches that relied on threshold models to grant pollution permits were unworkable for persistent toxic substances that mobilized into distant spaces and bioaccumulated up food chains.[15]

In its 1986 biennial report, the IJC acknowledged that great progress had been made on reducing conventional pollutants, such as phosphorus that caused algal blooms—but persistent toxics were not improving. "The visible conditions of Lakes Erie and Ontario, once considered shameful, have improved to such a remarkable extent that this achievement has been

internationally considered an unprecedented example of binational cooperation in environmental management." But "toxics were another matter. . . . Despite the positive signs of recovery from the severe stresses of
substances such as DDT, little overall progress has occurred in dealing
with the generic problem using technological means."[16] Words of praise
were followed by harsh admonishments. The 1986 report, for the first
time, stressed two core issues that would bedevil Great Lakes cleanup for
decades to come: polluted sediments and atmospheric deposition of contaminants. Neither could be easily addressed by simple limits on current
discharges. But cleaning up contaminated sediments required vast sums
of money, and preventing toxic inputs from other nations required creative
diplomacy and new international tools.

The 1986 IJC report was particularly notable for its sense of what historian Michael Egan calls "toxic fear." Toxics are now within us, the report
argued—and we have no idea how to understand them, predict them, or
avoid them: "We know that certain toxic chemicals pose grave risks to human populations. Intrusions of toxic chemicals on our lives, including the
incidence of specific toxic chemical disasters, are occurring with greater
impact and frequency. These occurrences cannot be viewed as system aberrations: rather, they may be the inevitable if unintended consequence of
our industrialized society. More toxic chemical-related incidents are bound
to occur. The questions thus become: How many? Where? How serious
will be the consequences?"[17]

Ecosystem approaches that had seemed like such an excellent idea in
1978 had become controversial by the mid-1980s because they were delaying actions on toxics. While paying lip service to the holistic ecosystem
approach, the 1986 report expressed concern about Environment Canada's
decision to eliminate funding for forensic research that searches for causal
links between wildlife injury and contaminant exposures.[18] Rather than focus on the interconnected web of multiple stressors that influence ecosystems, forensic approaches calculate which individual exposures have led to
specific effects. Forensic toxicology requires careful monitoring of specific
endpoints in wildlife health—but the Parties were eliminating those monitoring programs.

The commission urged "governments to reinstate remaining program elements that monitored the overall health of the herring gull population. . . . Herring gulls, which swim in the Great Lakes system, drink its
waters and eat its fish, are showing some signs of improved reproductive
capacity. Should this trend continue, it will be an encouraging development
since research shows that reproductive success in herring gulls is correlated
with chemical residues in the gulls and their eggs." But instead of expanding this program to other sentinel wildlife species, Environment Canada
decided to slash its funding. If you do not look for injury, it is easier to claim
that no injuries were found. Forensic toxicologists were so frustrated with
this process that when William Ashworth wrote *The Late, Great Lakes* in
1987, he argued that the promise of the Clean Water Act and Great Lakes
Water Quality Agreements in 1972 had been betrayed.[19]

What went wrong with ecosystem approaches? They had initially
sounded like an improvement on the narrow approaches that had led to the
aquatic nuisance program described in chapter 3. Of course ecosystems
are complex, and ameliorating pollution requires a broad understanding of
the entire context, not just a technical reduction in a single element. Yet, as
decades passed without much progress on further eliminations of persistent pollutants, critics, including ornithologist James Ludwig and ecologist
Michael Gilbertson, argued that the ecosystem approaches made banning
specific chemicals increasingly difficult.[20] When everything is interconnected, the political momentum needed to take action against one chemical
can easily fizzle. Industry can always call for more research to understand
the complexity of an ecosystem. Of course, calling for additional research
has long been a popular strategy for those opposed to regulation, but calls
for delaying regulation intensified when they gained the imprimatur of scientific authority from ecosystem approaches.

1987 Great Lakes Water Quality Agreement

Frustrated with the slow pace of toxics reductions, in 1987 the IJC amended
the Great Lakes Water Quality Agreement, insisting that the Parties create
specific timetables with detailed steps. The 1987 Agreement also required

the Parties to designate forty-three "Areas of Concern," or AOCs—the most severely polluted sites in the Great Lakes basin, such as Lower Green Bay and the Fox River in Wisconsin. The cleanups of these AOCs would be managed by teams of local stakeholders (called citizen action committees).

In a collaborative process, the citizen committees would figure out Remedial Action Plans (RAPs) that would lay out specific processes and timetables for restoring beneficial uses (fishing, recreation, and drinkable water). These stakeholder groups were given much of the responsibility for cleaning up local sites, yet they lacked adequate funding or staff support. Some RAP committees made little progress on contaminant cleanup. Others, however, including the St. Louis River RAP in Duluth, partnered with tribes and watershed groups, bringing together enthusiastic citizens who pulled on their rubber boots, tromped through contaminated estuaries, measured tumors on fish, planted endless trays of plants to help to remediate contaminated sites, and eventually pressured governments to fund effective cleanup projects.[21]

The 1987 Agreement required the Parties to create Lakewide Management Plans (LaMPs), which were documents that examined all the many stressors on each Great Lake, and then prioritized restoration needs and coordinated cleanup efforts. LaMPs had the potential to focus broad ecosystem approaches on the most important tasks for restoration. Yet instead, while the LaMPs collated abundant research, they also diluted the Great Lakes Water Quality Agreement's focus on persistent pollutants. For example, the author of one LaMP document stated: "The group of government agencies designing the LaMP felt it was also an opportunity to address other equally important issues in the Lake basin. Therefore, the LaMPs go beyond the requirement of a LaMP for critical pollutants, and use an ecosystem approach, integrating environmental protection and natural resource management." Even as the IJC urged the Parties to continue examining specific injuries to wildlife caused by contaminants, including fish tumors and reproductive failure in fish-eating mink, the Parties embraced the "interconnectedness" of ecosystems. James Ludwig, an ornithologist who did groundbreaking research on toxic contamination in Great Lakes birds, noted that when the ecosystem approach began to dominate Great Lakes cleanups, "fisheries researchers were considering the various stresses that could change the status of fish populations. These stresses included over-fishing, the introduction of exotic species and eutrophication. They concluded that they could not distinguish the specific cause of changes in the status of fish stocks. . . . To fisheries scientists, the specific cause tends to be not only unknown but unknowable."[22]

The Lakewide Management Plans that eventually resulted from the 1987 Agreement are fascinating documents, filled with interesting ecological and watershed information. Yet they have fallen short on specific policy recommendations and taken far longer to complete than predicted. The authors of the 2000 LaMP for Lake Superior admitted: "The LaMP process has proven to be a resource intensive effort and has taken much longer than expected. As a result, the public has had to wait years for a document to review."[23]

Remedial action plans, like the Lakewide Area Management Plans, were meant to embody the ecosystem approach, bringing together a wide array of diverse voices and perspectives and calling "upon the interactive talents available in a wide array of programs, including the involvement of local communities, citizens, and a wide range of organizations and government agencies." As one leader enthused, "Such groups provide an opportunity to change the traditional way of doing business by increasing the level of interaction." Not only that, these new groups would create new, experimental, "flexible institutional arrangements necessary to implement locally-designed ecosystem approaches to cleaning up degraded areas in the Great Lakes."[24] Scholars argued that these innovative approaches were wonderful examples of devolution. The countries were moving away from heavy-handed, top-down, bureaucratic command and control regulation (all very bad words in the new environmental discourse) and toward community-based restoration.

Not everyone was convinced, however. Scientists and negotiators within the IJC were concerned that community-based, stakeholder approaches might merely deflect responsibility and funding for cleaning up contaminated sites. A few years later, when progress had stalled, the IJC biennial report sternly warned of the dangers of deregulation and voluntary approaches, arguing that the history of Great Lakes pollution "teaches us that a strong governmental presence is indispensable to achieve progress on environmental issues. To sustain and build on these achievements, a strong regulatory base that sets common goals is needed. . . . Further progress and past accomplishments could be jeopardized if cutbacks in environmental legislation, regulation and funding for monitoring, enforcement and research are permitted to occur." The IJC warned that voluntary approaches "may be used to weaken and eliminate environmental laws" while reducing funding for essential monitoring and enforcement.[25]

The EPA responded to this IJC warning with a cheery assurance that "the U.S. will strive to ensure that any reductions in government funding will be offset by increased efficiencies in government programs and through increased participation from all Great Lakes stakeholders."[26] But it soon became clear that, without funding and government support, the citizen committees were stalled in their efforts to clean up the Areas of Concern.

Endless meetings and close ties to corporate parties in their local communities meant that many citizen groups became unwilling to require that industry clean up contaminated sediments, fearful that those companies would just pull out of town instead. Instead, they could all agree that multiple stressors were involved so no one need be held accountable. For example, one advocate of the ecosystem approach to cleaning up the Great Lakes argued that rather than institute bans, society first needed to undergo profound changes: "To implement an ecosystem approach and prevent adverse effects of toxic substances will require profound behavioral changes within society. Society must become less reliant upon land/water disposal and proactively pursue waste reduction, recycling, and use of destruction/detoxification technologies."[27] All these things may be excellent ideas. But saying they need to happen first is simply another way of saying that industry doesn't need to pay for sediment cleanup or reduce toxic discharges.

Zero Discharge Demonstration Program for Lake Superior: High Hopes Dashed

By 1989, many members of the public were deeply frustrated by the stalled progress toward cleaner Great Lakes. That year, at the IJC's fifth biennial meeting, the commission invited the public to testify, and hundreds of citizens and scientists spoke for hours on the "growing threat of persistent toxic substances and decried the lack of progress."[28]

The next IJC biennial report responded by calling for two key things. First, a precautionary approach, which meant "where approval is sought for the manufacture or discharge of any material or substance that will or may enter the environment, it should not be necessary for anyone to prove that the discharge will or may be harmful, but should be necessary for the applicant to prove that the discharge will not be harmful." And second, because of Lake Superior's "outstanding character and the unique threats posed by persistent toxic substances," the International Joint Commission challenged the governments of the United States and Canada to "make the Lake Superior Basin a zero discharge demonstration zone for all point sources of persistent toxic substances." The governments agreed, creating the Lake Superior Zero Discharge Demonstration Program to achieve an end to all discharges of persistent toxics into the Lake Superior Basin— and the eventual removal of legacy contaminants. Environmentalists were jubilant, calling this program a "toxic freeze strategy."[29]

Rather than add the Zero Discharge Demonstration Program to the responsibilities of existing federal agencies, Canada and the United States created the Binational Program, administered by federal, provincial, state, and tribal agencies. The hope was that, by bringing together federal, regional, and tribal governments, the Binational Program would not get bogged down in jurisdictional disputes. To increase citizen involvement, the governments also created the Lake Superior Binational Forum, a volunteer stakeholder group comprising members with diverse backgrounds that represent a wide range of perspectives—local government, industry, and business; labor, academic, and faith communities; recreation, environmental, and Indigenous communities. The forum's purpose was to ensure the success of zero discharge by fostering increased participation, acting as a watchdog group, and screening proposed developments such as new mines to make sure that they didn't discharge persistent toxics into the lake. Having served on the Binational Forum for half a decade, I can attest that these were hopeful, heady days for environmentalists and forum members.

But as environmental scientist and scholar Dave Dempsey warned, the flaws of the Zero Discharge Demonstration Program were written into the agreement: the document "did not set a final deadline for attaining the goal of zero discharge," and it allowed the two countries to agree to actions that were already underway, rather than new initiatives.[30]

These flaws became evident in the next several IJC biennial reports on progress and the responses by the Parties. In 1993, the IJC urged both Parties to prohibit all new point sources of bioaccumulative, persistent toxics into Lake Superior, as specified in the Zero Discharge Demonstration Program agreement. Canada simply refused.[31] The IJC responded with anger, pointing out that "zero discharge does not mean simply less than detectable. It does not mean the use of controls based on best available technology or best management practices that continue to allow some release of persistent toxic substances. . . . Zero discharges means no discharge or nil input of persistent toxic substances."[32]

Mercury was a particular concern, especially after the Grassy Narrows tragedy emerged (see chapter 3). The IJC urged an immediate stop to mercury discharges from chlor-alkali plants, such as the ones at Grassy Narrows. Canada responded by stating that it would "discourage" mercury use through voluntary programs but would not eliminate its release into Lake Superior. Canada's last chlor-alkali plant didn't close until 2008, fifteen years after this assurance that voluntary programs were leading to their closure.[33]

The pattern continued with mercury from smelter emissions, such as the sintering plant at Wawa on the northern shore of Lake Superior, a plant that caused extensive forest dieback and mercury releases. In 1991, the IJC called for an immediate end to mercury releases from mining and smelting. Canada responded by stating that it would suggest to a committee that it might consider looking into recommendations for reduction of mercury emissions from smelters.[34]

A few years later, the United States decided that the best way to eliminate its problematic mercury releases into Lake Superior would be by redefining dangerous levels of mercury—not by reducing mercury. The United States proposed to increase the "safe" daily intake level of mercury by a factor of five, allowing it to declare dramatic reductions in dangerous mercury levels. At the same time, however, the United Nations was consulting on research that led to a call for a fourfold decrease, not a fivefold increase, in the safe daily dose. As Visser and Ludwig argue, the same process happened with toxaphene monitoring and advisories.[35]

When the two Parties refused to implement immediate bans on toxic discharges, the IJC urged them to start by setting timetables for zero discharge. Environment Canada responded with a firm no: "The federal government does not support the recommendation. . . . Full account must also be taken of such factors as socio-economic impact." Canada continued, insisting that "it is premature to set a date for zero discharge. It should be remembered that the Lake Superior initiative is a pilot program. . . . It is a complex initiative requiring the continued commitment of many stakeholders. Achieving the program goal will not be a simple task."[36] Stakeholders and ecosystem complexities, in other words, had become an excuse — not just to refuse upholding their commitments to zero discharge, but to refuse even the basic task of setting timetables for upholding their commitments.

Every IJC biennial report, and response by the Parties, released between 1993 and 2003 shows the same pattern. First, the IJC criticizes the Parties for failing to eliminate toxic discharges — or even set timetables. Next the Parties give excuses for failing to do so. For example, in 1994 the IJC pointed out that "the governments have not taken a single concrete action to implement their zero discharge promise." When the 1994 IJC report called on the United States to "establish a coordinated, planned phase-out of existing sources" of dioxin, the United States insisted that it would not require the paper and pulp industry to eliminate dioxin or even to install "the best technology in process and treatment." Instead, the EPA would challenge industry to "voluntarily reduce generation of toxic substances." The United States added that setting a specific date for toxic elimination "may, in fact, detract" from voluntary agreements by industry. The EPA insisted that progress was being made because one Wisconsin paper mill had agreed to do yet another study on its waste streams to determine how it might reduce toxic releases. These studies had been going on since the 1930s with little progress toward elimination (see chapter 3). That same

year, Canada also refused to "adopt a specific time for the virtual elimination of toxic substances in the Lake Superior basin." Canada used the same
logic as the United States: specific timetables might interfere with "consensus amongst the stakeholders around the Lake Superior basin."[37]

Even partial agreements to discourage dioxin releases from pulp and
paper plants were reversed when industry balked. In 1996, after pressure
from the Ontario Forest Industries Association, the Canadian Ministry of
the Environment announced that it would "revoke a rule requiring pulp
mills to halt the discharge of chlorine-related compounds by 2002" and
also revoke a requirement that "mills explain how they would reach a zero
discharge goal." With growing frustration, the IJC report of 1996 accused
the Parties of using policymaking to keep toxics in circulation rather than
remove them.[38]

Scientific understandings of the complexity of contaminant transport
had led to two key challenges for Lake Superior cleanups. First, by the
early 1990s researchers were arguing that the majority of persistent toxics, such as mercury or toxaphene, in Lake Superior came not from local
industries but from atmospheric transport. Much of the mercury deposited
within Lake Superior, for example, came from coal-burning power plants
in the lower Great Lakes, along with some from China. How could local
industry actions reduce those sources? The second key concern was the
remobilization of legacy contaminants when sediments were disturbed. If
toxins from past industries didn't stay buried in the lake's sediments, did
that mean local communities had to find funding to dredge up all those
contaminated sediments?

The IJC feared that the existence of these complex sediment and atmospheric sources would become an excuse for polluting local industries
to continue polluting. And this is exactly what happened. In 1994, after
the IJC urged Canada and the United States to set timetables for industry
to stop releasing the "dirty dozen" into Lake Superior, the two countries
refused once again, stating "that they prefer to wait for more investigation,
including more information on atmospheric deposition, before making a
commitment on timing." Because ecosystems are complex, and because
toxics enter via the atmosphere and legacy sediments, the Parties used

these observations as an excuse not to regulate known discharges. The IJC objected strenuously, pointing out that "the fact that there are inputs other than intra-basin point sources should not prevent or delay action on those specific sources."[39] Yet that is exactly what it did.

Two years later, the IJC suggested that North American companies had shifted pollutants from water discharge to atmospheric discharge in order to avoid regulation. The IJC stated: "A recent report of the Canadian Chemical Producers' Association, 'Reducing Emissions,' illustrates that from 1992 to 1994 total Canadian water emissions by member companies were halved, but the proportion of air emissions almost doubled."[40] Within Ontario, the IJC noted, nearly 99 percent of emissions from members of the Canadian Chemical Producers' Association were now directed into the air rather than the water. It was a bitter pill for the IJC to swallow: North American industries were insisting that new research on atmospheric transport meant that it was unfair to insist that industry needed to reduce toxics in water effluent because most toxics in Lake Superior came from the air. But they failed to mention that many of those airborne toxics came from their efforts to avoid pollution controls, not from growing economies in the third world.

The 1998 IJC report opened with a plaintive query, "Why are we unable to effectively deal with these persistent toxic substances?" The report went on to discuss its "introspective search" that led the IJC to ask whether "there is a purpose for the biennial reports" given that progress was "overshadowed" by "the immensity of the work still ahead" to clean up the Great Lakes and the fact that federal, state, and provincial governments on both sides of the border seem to "have forgotten their responsibility to protect the citizens they represent."[41] With this glum report, the IJC essentially gave up insisting that the Parties comply with the Great Lakes Water Quality Agreements. In recognizing that Great Lakes governance was shifting from "command-and-control emphasis to voluntary measures," the IJC also seemed to give up its responsibility to serve as a watchdog for the Great Lakes.[42]

After the 2002 biennial report, the IJC largely stopped discussing toxics — even though toxics were still supposed to be the focus of the Great

Lakes Water Quality Agreement. Toxics hadn't been cleaned up, by any means, but they had just become too depressing. The 2004, 2006, 2009, and 2011 biennial reports all largely ignored toxics, focusing instead on invasive species, pathogens in groundwater, and sanitary hygiene.[43]

The 2013 report discussed toxics, but it spun the results as largely positive: "Since 1987, all seven indicators of chemical integrity have shown mostly favorable or stable results. The levels of many persistent toxic chemicals entering the Great Lakes from atmospheric deposition are lower than they were in 1987. Concentrations of most measured persistent toxic chemicals decreased in herring gulls, fish, sediments and mussels. Most reductions occurred from 1987 to 2000, but since 2000 trends vary by chemical, location, and species."[44]

When you look at the actual data, however, the results are much less cheerful. After nearly $1.5 billion spent in the Great Lakes Restoration Initiative to get measurable results in cleaning up AOCs, progress was still slow. "Beneficial use impairments" include "loss of fish habitat or contaminants in fish serious enough to prompt consumption warnings." Across the forty-three Areas of Concern in the Great Lakes, less than 10 percent were cleaned up and delisted in twenty-five years. About a fifth of the awkwardly termed "beneficial use impairments" were improved in that same time. At that rate, it would take at least another century to clean up the toxic detritus from a few postwar decades of industrial enthusiasm.[45]

Dave Dempsey, long involved with the IJC in its efforts to ensure the implementation of the Great Lakes Water Quality Agreements, argues that they were ultimately ineffective because of a "reluctance by politicians to break ranks with industrial interests defending the status quo."[46] Currently, enormous political pressure exists to delist Areas of Concern, even if they're really not cleaned up. As a Lake Superior Binational Forum member, I was involved in evaluating the EPA's proposal to remove Deer Lake from the list of Great Lakes Areas of Concern. Deer Lake had been contaminated with mercury from iron mining because the Cleveland-Cliffs Iron Company discharged mercury-containing wastewater into the city of Ishpeming's wastewater treatment system between 1929 and 1981. From there it moved into Carp Creek and made its way into Deer Lake, where it accumulated in fish. In 1981, after high concentrations of mercury were found in Deer Lake fish, the Michigan Department of Community Health put a ban on eating all fish from Deer Lake. Six years later, Deer Lake was listed as an Area of Concern. Three main problems (the so-called beneficial use impairments) affected the lake: eutrophication leading to excessive algae growth; wildlife health problems, including deformities and failed reproduction; and mercury accumulation resulting in fish consumption advisories. After a twenty-seven-year effort with multiple stakeholders from the local community, and grants totaling $8 million from the Great Lakes Restoration Initiative funds, the community of Ishpeming was able to divert Partridge Creek from flowing through the old mine site, helping to significantly reduce (but not eliminate) the final controllable source of mercury contamination in Deer Lake.

Yet even after this effort mercury levels in Deer Lake fish remained too high for people to eat the fish. Yes, mercury levels had decreased from the very high levels of the 1980s, but they were still higher than in Lake Superior and nearby inland lakes. The downward mercury trends observed in the early years had stabilized at levels too high for safe fish consumption. This meant that the Michigan Department of Community Health still had a fish consumption advisory on fish from Deer Lake because of excessive mercury levels, some from atmospheric deposition and some from the history of mining contamination. Yet with delisting, most people will assume this means that fish are now safe to eat. In fact, on June 18, 2014, Mark Loomis—the task force leader from the EPA's Great Lakes National Program Office—was quoted in the local paper as saying: "When you go out there now it's clear water and the fish are safe to eat." A great deal of effort will be required to educate the public that the delisting of the Area of Concern does not mean that mercury levels have declined enough to make the fish safe to eat.[47]

Emerging Contaminants: Flame Retardants Rising in the Great Lakes

Flame retardant chemicals illustrate continuing challenges facing the Great Lakes. Like PCBs, dioxins, toxaphene, and mercury, flame retardants persist in the environment. They can accumulate in fat, magnify their way up food chains, have dramatic effects on thyroid hormones, and cause cancer and birth defects. These attributes mean that they mobilize into distant places, insinuating themselves into the most intimate of bodily spaces. Like PCBs, flame retardants were designed to resist being broken down in the environment—precisely what made them effective insulators and flame suppressants. But it is also the quality that makes them harmful. Unlike those other chemicals, most flame retardants are still in widespread use.[48]

Flame retardants such as the polybrominated diphenyl ethers (PBDEs) are chemicals added to plastics, foam, and textiles. Up to 35 percent of these items by weight consist of flame retardant chemicals. Because they are not chemically bound to the materials, they easily migrate into the broader environment. When the foam begins to crumble inside your couch, when your child's ketchup-covered clothing gets tossed into the washing machine, the chemicals slip into the air, dust, and water, beginning their journey round the world. If they stick to dust particles, they can become airborne and ride atmospheric currents into the north. Or they can stick like a burr onto soil and wash into the rivers and from there into lakes where they may attach to bottom sediments, until carp and other bottom-dwelling fish remobilize them. Once they've hopped into global circulation, many of them resist being broken down into nontoxic molecules.[49]

Chemical flame retardants were first used by the Romans, but they did not become popular until the post–World War II synthetics boom. After the war, plastics, foams, and synthetic fibers were widely promoted as the modern, sanitary future of the American home. Consumers turned away from traditional wood and metal furnishings and plant-based fabrics and began to choose the shiny, plastic appliances, sleek modern fabrics, and comfortable polyurethane foams in furniture. But there was a serious problem: these new synthetic materials were far more combustible than traditional materials. Smoking rates were also increasing, leading to a deadly combination—a rise in fires caused by smoking materials. Between 2009 and 2013, almost a quarter of deaths from domestic fires came from fires started by smoking, which persists as the leading cause of fire-related deaths.[50]

In the mid-1970s, the young Consumer Product Safety Commission required manufacturers to add flame-suppressing chemicals to many fabrics and upholstery foams and plastics. Rather than urge Americans not to smoke in bed or require cigarettes to resist ignition, the chemical industry borrowed a page from the tobacco company's playbook and heavily promoted flame retardant chemicals, particularly those that contained bromine or chlorine.[51]

Flame retardants, like PCBs, had been promoted as stable and safe. But by 1976, just a few years after they made their way onto the market, some researchers were warning that they might be more toxic than expected. In 1976, American researchers R. Liepins and E. M. Pierce noted that "only a few" of the plastic additives "have been evaluated in any great detail," yet they were being promoted for consumer products. They noted that 384,000,000 pounds of flame retardants were already being used in American products, and they predicted growth of 17 to 20 percent each year. Yet remarkably little was known about their toxicology, Liepins and Pierce admitted.[52]

Five years later, Swedish researchers detected the flame retardants named PBDEs in fish caught in a river on the west coast of Sweden. This was puzzling because PBDEs were consumed at very low rates in Scandinavia compared to the United States. Other researchers began looking for the chemicals far from their sources of production and consumption. Levels were astonishingly high in marine and river sediments across the world. Marine shellfish and fish in Japan contained them, as did cod livers in the North Sea. More Swedish studies found them concentrating their way up the food chain into grey seal and seabird eggs. By 1998, they had reached the deep ocean and were present in sperm whales, which normally stay and feed in deep water. Remarkably high concentrations were found in wild salmon from the Baltic Sea and in harbor seal blubber.[53]

What about people? Were PBDEs making their way into humans? In 1998, the Swedish researcher Per Ola Darnerud detected flame retardants in the Swedish food supply, including fish, milk, and eggs. The next year, other Swedish researchers reported a sixtyfold increase in flame retardant concentrations in women's breast milk sampled between 1972 and 1997.

Concentrations were doubling every five years. Very high levels were also present in the blood of Swedish workers in electrical manufacturing—not surprising, because they were used in insulating materials. But surprisingly high levels were also present in the clerks who sat in front of computer screens, suggesting that somehow they were making their way out of the computers and into the dust and air. In 1999, Swedish researchers warned that flame retardants might act as hormone disruptors, similar to the most potent PCBs in their effects on the thyroid. After the 1999 report from Sweden, researchers looked and found them in human tissue in places as far-flung as Japan, Israel, and Spain. Samples from lake trout collected in the Great Lakes in 1997 also showed very high levels of flame retardants.[54] Soon after these reports, Europe moved to ban the materials, afraid they were about to have another PCB tragedy on their hands. North American regulators, in contrast, continued to promote them as the answer to fires, necessary for children's health.

European Union researchers were puzzled by the North American refusal to act. They agreed with American researchers that the animal data on PBDE toxicity were incomplete and that little was known about human effects. But as Ilonka A. T. M. Meerts of Wageningen University and Research Center in the Netherlands said, "The complete toxic profile is very much like PCBs."[55] Most important, while there was no single line of direct evidence that they killed people, concerns from multiple lines of research converged to warrant caution.

After flame retardants were banned in Europe, their use continued to rise in United States and Canada. In the United States, in 2003, researchers reported very high levels in breast milk. Concentrations were as high as 419 parts per billion lipid in one study: 10–100 times greater than human tissue levels in Europe. Researchers noted, "Their detection in breast milk raises concern for potential toxicity to nursing infants, given the persistence and bioaccumulative nature of some of the PBDE congeners."[56]

A year later, Canadians had reason to worry as well. The *Globe and Mail* in 2004 reported, "The breast milk of Canadian women contains the second-highest levels in the world of a compound used as a flame retardant in computer casings and household furniture." American mothers had the highest levels, but Canadian mothers also quite high: with levels five to ten times those in Europe, "exceptionally high compared to those elsewhere in the world." Unlike the banned POPs, such as DDT and PCBs, whose levels had fallen after their bans, levels of flame retardants had been growing exponentially in human milk. By 2004 in North America, they were one hundred times higher than international samples from the 1970s.[57]

Rather than respond to these findings, Canadian regulators followed the lead of American regulators, deciding to wait for more research. Health Canada official Samuel Ben Rejeb, associate director of Canada's Bureau of Chemical Safety, said that the department was "studying why levels in Canadian women are so much higher than elsewhere in the world." They were indeed very "interested in PBDEs as a new emerging class of persistent organic pollutant." But their industry partners had assured them "human exposure from sources such as breast milk had not yet reached harmful levels." So, therefore, the Canadians decided there was no need to act with caution until conclusive proof of harm and injury had been received. "The highest reading in Canada was of one woman who had 956 parts per billion of PBDEs in the fat of her breast milk. The highest in the U.S. was just over 1,000 ppb. . . . PCB concentrations become of concern when they reach 1,250 ppb, according to Mr. Muir." Some of their government staff scientists were furious: Muir told the *Globe and Mail*, "PBDE levels in the breast milk of a small number of women surveyed in North America are approaching the critical concentrations associated with health impairment from PCBs. . . . This is a poster-child chemical for something that ought to be zeroed out."[58]

Women in the Great Lakes region continue to have some of the highest levels in the world of flame retardants because the Great Lakes are a sink for atmospheric transport of the chemicals. They mobilize from crumbling foam and burning plastic, into dust and the air, and from there they fall into the lakes and became bound, briefly, to lake sediments before bouncing back up into the lake food chains—and from there into the women who eat those fish. Flame retardants make their way from clothes and human bodies, into sewage systems, and from there into sludge that is spread onto fields. "The average concentration of [one flame retardant] in sludge in the Lake Superior watershed was 510 parts per million (ppm) and in the Lake Michigan watershed was 466 ppm." In those Lake Superior and Lake Michigan communities, PBDE congeners "were found to be 30–50 times higher than the most abundant congener of PCBs (polychlorinated biphenyls), which are chemically very similar to PBDEs."[59]

From the sediments, those flame retardants move into Great Lakes fish and the birds that eat those fish. One study by Zhu and Hites shows the levels of flame retardants rose exponentially between 1980 and 2000, increasing three-hundred-fold in Lake Ontario lake trout, with similar increases in Lake Michigan lake trout. Similar trends in lake trout have been reported by Luross et al. for Lakes Ontario, Huron, and Superior. In a 2001 study, Lake Michigan salmon contained PBDEs at levels above 100 parts per billion, "one of the world's highest concentrations for salmon in open water." Another study that same year showed a sixtyfold increase of PBDEs in herring gull eggs from the Great Lakes region from 1981 to 2000.[60]

When William Sonzogni, director of Wisconsin State Laboratory's Environmental Division, searched for PBDEs in Lake Michigan salmon, his team found PBDEs "in all the sampled fish at average concentrations of 80 ppb, six times higher than PBDE levels found in Baltic Sea salmon. Compared to other studies that have examined PBDE levels in fish . . . among the highest reported in the world for salmon in open waters." Soon afterwards, the state lab confirmed finding PBDEs in "blood samples drawn from five bald eagles inhabiting the southwest bays of Lake Superior near Duluth and Superior. Canadian studies of herring gull eggs from around the Great Lakes report rises in PBDE concentrations between 1981 and 2000."[61]

And not surprisingly, the people who eat those Great Lakes fish also had high levels. For example, in one 2009 study of health professionals from across the country, every single participant contained some PBDEs inside his or her body. But the two people from Michigan in the study have particularly high concentrations — and a wide variety of different flame retardants in their blood. One Michigan woman had particularly high levels of PBDEs in her breast milk. As the *Chicago Tribune* reported, "A typical American baby is born with the highest recorded concentrations of flame retardants among infants in the world"—and levels in Great Lakes babies are among the highest in America.[62]

In 2006, the Lake Superior Binational Program indicated "concern" about these emerging contaminants, noting that "while concentrations of PCBs, DDT, and other 'legacy' pollutants have been declining in the Lake Superior environment, analysis of archived Lake Superior whole lake trout tissues shows that PBDE concentrations increased exponentially between 1980 and 2000 with a doubling time every 2.5–3 years." But the report also insisted that because monitoring data are limited, "an evaluation of their environmental presence and potential effects to the Lake Superior ecosystem" would be "difficult at this time."[63]

Laboratory animal studies have found PBDEs to cause learning, memory, reproductive, and liver problems as well as cancer. They are toxic to developing brains. Above all, they are potent endocrine disruptors, disrupting thyroid hormones. But as with all other endocrine disruptors, it is nearly impossible to point to definitive evidence that they kill people in low doses. Linda Birnbaum, director of the Human Studies Division at EPA's National Health and Environmental Effects Research Lab, told reporters in 2001, "We banned the production of PCBs when we had less information than we do now of the PBDEs." She warned that the similarities were profound: both PCBs and PBDEs had key structural similarities to thyroid hormones. They all consisted "of two six-carbon rings decorated with halogens." Like PCBs metabolites, they bind transthyretin, a chemical evolved to transport such thyroid hormones as T_4 to the brain of the developing fetus. This, to put it mildly, would not be good for the child, since it would interfere with neural development. In people and animals, thyroid hormones have powerful roles in the developing brain and nervous system. As early as 2001, numerous studies had found "that rodents fed high amounts of PBDEs have thyroid hormone deficiencies," reducing T_4 by about 70 percent in one study. Birnbaum said: "This is not a case where we are looking for missing arms and legs. . . . We're looking at reduced ability to learn, altered behaviors, decreased sperm count, premature ovarian failure—things that are more difficult to pick up in the standard studies."[64]

Global use of these chemicals soared, from 526 million pounds in 1983 to about 4.4 billion pounds in 2014. Why did these chemicals persist on the market? Not because they've been shown to reduce fire risk or death from fire risk. The *Chicago Tribune* in 2012 had an excellent investigative series on science, industry advocacy, and flame retardants. Even as "evidence of the health risks associated with these chemicals piled up, the industry mounted a misleading campaign to fuel demand." The *Tribune* series discusses what they call a "decades-long campaign of deception that has loaded the furniture and electronics in American homes with pounds of toxic chemicals linked to cancer, neurological deficits, developmental problems and impaired fertility. The tactics started with Big Tobacco, which wanted to shift the focus away from cigarettes as the cause of fire deaths, and continued as chemical companies worked to preserve a lucrative market for their products, according to a *Tribune* review of thousands of government, scientific and internal industry documents."[65]

Lessons Learned from Great Lakes Toxics

What can we learn from history to help us address emerging contaminants like flame retardants? As we are now faced with many billions of dollars of cleanup costs from long-banned chemicals in the Great Lakes, it makes little sense to introduce new ones with similar modes of action. The strategies now being used to delay regulations of flame retardants are familiar from debates about tobacco, endocrine disruptors, and greenhouse gas emissions. By casting doubt on research and calling for more studies, industries can delay regulation for decades, even when they know a substance causes injury. Requiring proof of harm from exposed communities only delays effective action. Because the history of pollution in the Great Lakes is one of late reactions to growing evidence of injuries, many scientists have agreed that precaution is critical. The burden of proof should be on the producer, not the consumer.[66]

The Toxic Substances Control Act (TSCA) of 1976 is part of the problem. Passed after a great deal of lobbying and rewriting by chemical companies, TSCA places the burden of proof on the exposed communities and regulators, meaning that it becomes nearly impossible for the EPA to ban dangerous chemicals. Under TSCA, manufacturers are not required to submit safety data, which has forced the EPA to predict possible health problems with computer models that often fail to identify known problems. Moreover, to ban a chemical that is on the market, the EPA must "prove that it poses an 'unreasonable risk.' Federal courts have established such a narrow definition of 'unreasonable' that the government couldn't even ban asbestos." Trade secret rules mean it's almost impossible to track certain chemical movements, much less their potential injury.[67]

The EPA has been turned into a pollution-permitting agency, deciding how best to allow pollutants onto the market—not a pollution-control agency, as many people assume. Intended to monitor risks, EPA administrators are profoundly risk-averse about threats to their own agency that might arise from trying to protect public health. It is always safer to call for more research: "We are always learning," said Jim Jones, the EPA's acting assistant administrator for chemical safety and pollution prevention. "We want to make sure we have a better understanding of the human health and ecological risks before we commit to any course of action." Such research takes time, which keeps harmful chemicals on the market for many years. Since the Reagan antiregulatory era, an average of twenty-five years typically passes between the time a chemical is found to cause injury and "the appropriate and sufficient response from the authorities."[68]

The focus on ecosystem interconnections appeals to many researchers, but it slows effective regulation. There is always another stressor to model, another interconnection to study, which means that trying to ban any particular contaminant is rarely worth the political effort. Michael Gilbertson, an environmental scientist who worked for decades on Great Lakes toxics, writes that this "supposed complexity and uncertainty has not been inconvenient to those interests reluctant to implement the costly remedial policies contained in the Great Lakes Water Quality Agreement."[69]

Canadian and American federal regulators have turned to voluntary agreements with industry to phase out certain chemicals that pose potential risks. But voluntary agreements may actually lead to increased exposure—just in a different place—rather than reduced exposure. The saga of a particular flame retardant called chlorinated tris, retold in the *Chicago Tribune*, is an excellent example. After it was linked to cancer, manufacturers voluntarily took chlorinated tris out of children's pajamas. "But because it wasn't banned, companies could legally use it in other consumer products without informing government officials or the public. . . . In a statement, the EPA said it is largely powerless to do anything about chlorinated tris."[70]

Voluntary approaches in Lake Superior have often led to more exposure to mercury, more exposure to dioxins, more atmospheric releases. Exposed communities such as the Anishinaabe in Grassy Narrows didn't volunteer to be exposed to mercury, or to the collapse of their fishing economy. A voluntary approach is hardly voluntary for exposed communities.

Climate Change, Contaminants, and the Future of Lake Superior

IT'S COLD OUT TONIGHT, UP HERE in the north woods of Wisconsin. Wind chill warnings are going out over the weather networks, reminding us that the 40 below zero wind chills predicted for the evening could freeze exposed skin in ten minutes. "If you have to venture outside," the radio voice tells us, "cover all extremities. Carry emergency gear. Stay home if possible."

On a night like this, with the wind whipping off Lake Superior, thrusting snow eddies into the cabin every time you crack the door for more wood, it's hard to believe in global warming. In the village's general store, people stand around the pellet stove stomping feeling back into their toes and joking about how we could use a little of that global warming right about now. Thaw out those frozen septic systems, maybe. Cut down on those heating bills. Maybe speed up spring a little while, so it doesn't start on July 5.

But underneath the jokes, they all know in their hearts that something big is changing. The north is no longer as cold, as extreme, as bitter and brutal. To some people, luxuriating down south on their golf courses and cruise ships, it might seem like no great loss. But to those of us who live in the northern forest, we know it *is* a great loss. John Burns, a naturalist in the North Woods of Wisconsin, writes in *Paradise Lost? Climate Change in the North Woods:* "The climate change scenarios currently projected for Wisconsin at the end of this century utterly boggle the mind. Conservative middle-ground scenarios show Wisconsin becoming the climatological equivalent of Arkansas, while Madison's climate will morph into a twin of Oklahoma City. . . . Meanwhile, the North Woods may gradually transition into an oak savannah. That's so difficult to imagine, so close to what we can only think of as science fiction, that all of us have a great deal of trouble even conceiving of the possibility. Yet there it is, looming on the horizon like the eerie bruised sky that so often precedes a tornado. But how does one address the coming of a tornado, much less the coming of a global environmental upheaval?"[1]

The potential loss is difficult to comprehend—not just because eagles, lake trout, wolves, spruce, and birch may all decline, and not just because of the smoke that will fill our lungs from fires, or the mercury that will be released into the atmosphere from those fires. Ultimately, all those losses pale compared to what we might also lose: a small but significant part of our own human selves.

What is often missing in discussions about global warming is some sense of the human dimensions. This is puzzling, because put bluntly, global warming is a humanistic tragedy, not an environmental tragedy. It's not the earth that's going to be devastated by climate change. Mass extinctions in evolutionary history are typically followed by mass speciations. New species will surely evolve to fill the niches vacated by the extinctions piling up in ever greater numbers. The earth will persist—different, surely, but that's the way things go in evolutionary time. The looming terrors of global warming—massive fires, floods, famines, with refugees in the millions fleeing too much water in Bangladesh, too little water in the Sudan—are tragedies in historical time, not in evolutionary time.

Lake Superior is one of the fastest warming lakes in the world, leading to enormous changes in its limnology, so that it may not remain a very cold, oligotrophic lake for much longer. For those of us who love the basin, it is hard to comprehend—even for people who don't know what oligotrophic means. You don't need to understand the details of limnology to understand something key is changing when Lake Superior begins to warm so quickly that algal blooms become common.

Climate change's effects on Lake Superior is an enormous topic, subject of thousands upon thousands of pages of recent research and policy discussion. This concluding chapter won't try to review all the material. Instead, the chapter will ask: How might changing climates affect the mobilization of contaminants in the Lake Superior basin? How might those contaminants affect resiliency toward climate change? And what can we do about it?

Weather and climate have enormous variability, independent of any human activities. Temperatures change over the course of a day, a season, a year, and over decades as well. To sort out patterns of longer-term change from the natural bumpiness in climate data, the natural variation, scientists can turn to historical records, which let us begin to see patterns across longer time periods.[2] The computer models known as General Circulation Models (GCMs) allow scientists to predict what might happen in the future under various levels of emissions. To help ground-truth these models, climate scientists compare their predictions with data found in a variety of historical archives, including paintings, books, museum specimens, ice cores, and sediments in lakes.

Because it is so vast, Lake Superior has an enormous influence on the region's climate, moderating temperatures so summers are cooler and winters are warmer, compared to places inland. Historical sources reveal that summers have been fairly mild in the past several centuries, with high temperatures that have averaged around 68 to 77°F (20 to 25°C). Winter lows have also been moderated by the lake, so that low temperatures in Duluth have averaged about 5 to -4°F (-15 to -20°C). In recent decades, temperatures in the basin have increased by 1°F, which may not sound like much but can create significant ecosystem changes. By the end of the century, projected temperatures will be from 5.4 to 8.1°F warmer. Winter temperatures will increase even more sharply, with models predicting increases of up to 11°F. By 2080–2099, our cool summer climate in the Upper Peninsula of Michigan will have become much more like the climate of southwestern Kansas.[3]

The current annual average water temperature of Lake Superior has averaged approximately 40°F—which is extremely cold for a big lake. But the surface warms much more quickly, and by September, surface water temperatures can reach 59 to 77°F. Because lower-density warm water floats above higher-density cold water, the warmer surface waters create summer stratification. This means that a layer of dense cold water separates from the warmer surface water, and for much of the year, those two layers do not mix. But twice a year, turnover happens. In the late fall, as air temperatures cool, the surface waters cool as well. Once surface temperatures reach 39°F, the water sinks and deep water rises, causing vertical mixing in the water column. In the spring, turnover happens once again, bringing oxygen from the surface down into the deeper waters, while also resuspending nutrients that had been trapped at the bottom.[4]

Water temperatures in Lake Superior are already rising at twice the rate of air temperatures in the basin — one of the fastest rates of change in the world for large lakes. Research by Jay Austin at the University of Minnesota–Duluth shows that surface water temperature in summer increased between 1979 and 2006 by 4.5°F, or about 2°F per decade, while regional air temperature increased about 1°F per decade. At the eastern end of the lake, average water temperatures increased about 6.3°F during the century between 1906 and 2006, with most of the warming occurring after 1976.[5]

Observed ice cover on Lake Superior has dropped 79 percent in recent decades — 2 percent per year for the period 1973–2010. By February, the lake used to average 75 percent ice cover, but by the end of this century, we'll be lucky to get 10 percent ice cover (figure 9.1). Ice duration is also a lot shorter; it may drop by sixty days to less than half its current duration by the end of the century. Less ice cover for a shorter amount of time means the lake warms up earlier in the summer. By mid-July, in recent years, Austin's research shows, surface temperatures were 15 degrees warmer than normal. This, in turn, means earlier summer turnover and stratification. The surface waters that warm up have extended deeper into the lake, changing the behavior of fishes that prefer the deeper layers.[6]

Snow is key to the Lake Superior basin's identity and ecology (figure 9.2). My first winter in the Keweenaw Peninsula, which catches lake effect snows, we received 340 inches of snow. It was impressive but still well below the 390 inches that fell in 1978–1979.[7] One winter in the 1930s, snow fell every single day for two months solid.

Climate change in the basin may not change total annual precipitation, but less will fall as snow and more as rain. Summer storms are likely to be more intense, punctuated by drought. Soil moisture could decrease by

Figure 9.1 Ice cover on Lake Superior, February 16, 2014. (Satellite imagery courtesy NASA/MODIS: NOAA CoastWatch–Great Lakes Region.)

Figure 9.2 Railroad in winter, Calumet, Michigan, no date. Winter snow has shaped the Keweenaw Peninsula's ecology and identity in profound ways. (Roy Drier Photographic Collection, Michigan Technological University Archives Neg. 00223.)

30 percent in summer, affecting plant growth. Runoff may increase, leading to more erosion, more pesticide and nutrients from fields washing into the lake, and more sewer overflows that dump raw feces into the lake. Already, 1,260 billion gallons of untreated wastewater spill into the Great Lakes each year, and without expensive investments in new infrastructure, that is likely to increase.

The vast majority of the Lake Superior watershed's biodiversity cannot be seen, hiding within the soil or under the water's surface. Soil microbes influence carbon and nitrogen cycling, and climate influences soil microbes. Recent warmer autumns mean that there is less early snowfall, which insulates the soil from deep freezes. Without the snow, the soil may freeze more deeply, bursting microbes, which then release nitrogen and carbon into the soil. Nutrients washing from the soil move into streams and then the lake, where they can feed algal blooms and contribute to eutrophication.[8]

Adult whitefish and lake trout will not suddenly die as their water warms; they can swim into deeper offshore waters where the water is still cool. But their reproduction may be affected and their predators may increase. Whitefish have higher reproductive success when their eggs are protected by winter ice, for example, so declining ice cover may put pressure on their populations. Predators such as sea lamprey feed faster, grow bigger, and lay more eggs in warmer water, which means they may become more of a threat to lake trout again.

Carp, zebra mussels, and many other invasive species will expand their ranges. Currently, Lake Superior contains a complex mix of native and nonnative fish. Some nonnatives snuck in on their own, including ruffe, round goby, and the sea lamprey. Others were intentionally introduced to make anglers happy — the Chinook salmon, Coho salmon, brown trout, and a variety of rainbow trout strains. As water temperature warms, fisheries scientists fear that cold-water fish species, including lake trout, brown trout, and Coho and Chinook salmon will decrease, while warm-water species may increase. Zebra and quagga mussels in the lower Great Lakes already clog water intake pipes and devastate native freshwater mussels. So far, Lake Superior's cold water has limited the northward expansion of zebra and quagga mussels, but warming temperatures may allow them to flourish.[9]

Forests will change as well, with models predicting that our forests may become similar to those now in Arkansas. Nearly 85 percent of the Lake Superior basin is currently forested, with a mixture of boreal forests in the north and aspen-birch, spruce-fir, maple-yellow birch, and white-red-jack pine trees along the southern shores (figure 9.3). Climate change means that existing stresses on trees will increase, including drought, wind, fires, and insects, leading to large-scale tree mortality, reduced regeneration of trees from increased deer herbivory, and loss of boreal forests. With all that dead wood, forest fires may increase.[10]

Summer drought means that boreal forests along the north shore of Lake Superior will vanish if continued warmer air degrades habitat for cold-loving trees like paper birch, white spruce, balsam fir and white ce-

Figure 9.3 Old white pine along the Lake Superior shoreline, Bete Grise Preserve, Michigan.

dar. Boreal regions contain just under 15 percent of the global land surface, but they contain over 30 percent of all terrestrial carbon, stored largely in the soil. Low soil temperatures have historically increased the formation of peat, and low decomposition rates in the cold temperature have promoted high rates of carbon sequestration. As temperatures rise, carbon may decompose faster, meaning that one of the key carbon sinks of the world might tip into a carbon source instead.[11]

Climate and Contaminants

Climate change will affect the ways that global contaminants are transported and mobilized within Lake Superior food webs. Recall from chapter 3 that transport pathways help determine the risk of contaminants. If a contaminant from distant shores is going to get into the fish that swim beneath my rocks, and from that fish into my body and cause harm, that contaminant must first be transported by ocean currents, rivers, or migratory species before it is metabolized into something less harmful.[12] An extremely toxic poison presents little risk if it is buried securely beneath miles of ice, while a less toxic poison, such as alcohol, can become risky when we're exposed to large quantities faster than we can break it down.

Climate change will have a powerful effect on the environmental fate and behavior of contaminants because it will change the drivers that determine how chemicals partition themselves into air, water, soil, and biota. Temperature determines the rates of key processes such as air-surface exchange, evaporation, wet and dry deposition, metabolism, and degradation. Climate change will alter all these metabolic, deposition, and transport processes, but in surprising and unpredictable ways.[13]

Semi-volatile compounds such as toxaphene and many PCBs bounce their way north like grasshoppers. Warming temperatures are likely to increase the rate of cycling of these semi-volatile chemicals, so the grasshoppers will move farther and faster, bouncing up into the atmosphere and then back down into northern waters at higher rates. Decreasing ice cover will also increase contaminant mobilization for these semi-volatile compounds. When a lake is covered by ice, those chemicals stay put in cold, icy waters. But when the ice melts and the surface waters warm, up they go, zipping along farther north. This might allow toxaphene and PCB levels to drop in Lake Superior fish; instead they will rise in fish that swim in waters to our north.[14]

Another key avenue for exposure is migration. When fish, birds, and caribou migrate, they become vectors for contamination, bringing chemicals borne in their body fat with them as they cross oceans and move up rivers and streams. For example, American eels migrating from Lake Ontario contaminate beluga whales in the St. Lawrence estuary with persistent organic pollutants. Birds in the Arctic bring contaminants from oceans to terrestrial systems via their guano. Researchers in Canada found that in ponds under the nesting cliffs of northern fulmars, a long-distance migratory seabird, DDT and mercury concentrations were ten to sixty times higher than expected. Levels of PCBs in sub-Arctic lakes with abundant salmon runs may be eight times higher than in similar lakes without migratory salmon. Because temperature influences fish migrations, warmer temperatures may increase the transport of contaminants, bringing more chemicals to the north.[15]

Migration doesn't just change the exposure of lakes to toxics; it can also change the effective toxicity of the contaminant for individual animals. An exhausted, starving animal has a suppressed immune system, increasing the hazard of a particular level of toxic chemical. Additionally, when a bird or a fish migrates a long distance, it fuels the journey by burning off much of its stored fat—releasing any toxics that had been stored in the fat back into circulation. When a female fish lays eggs with her dwindling energy stores, her concentrated toxics mobilize into the eggs, exposing her offspring to extraordinarily high levels.[16]

Increases in river temperature during fish migration may magnify the stress on migratory fish, leading to more susceptibility to the toxics being released from their fat. If this happens, thermal stress combined with contaminant mobilization may suppress reproduction. Sublethal concentrations of toxics that don't kill animals can alter their behavior in subtle but critical ways. Young salmon exposed to low levels of DDT shift their thermal preference, seeking out cooler water. But if climate change means that cooler water isn't available for them, what happens? Robie Macdonald warns, "It is effects such as these that hold the greatest potential for surprises when, for example, climate change produces altered thermal regimes in aquatic systems lightly impacted by endocrine-disrupting chemicals."[17]

Invasive species threaten the resilience of cold northern lakes. Many of the species that are moving northward come from warmer waters, which will "amplify their competitive advantage over the cold-water species of the Great Lakes as the temperature of waters increases." And as they invade, they may also increase the biomagnification and mobilization of contaminants. For example, zebra mussels suck up vast amounts of lake water, concentrating PCBs in their bodies. That would be an effective way of removing PCBs from open water—if we then gathered up zebra mussels and took them out of circulation. But instead, zebra mussels with their toxic bodily burdens become food for round gobies, which concentrate the PCBs even more, and then get eaten by walleyes, making these delicious fish, favored by generations of families at Friday night fish fries, too poisonous to safely eat.[18]

The toxicity of certain contaminants may also increase with temperature, rendering that chemical more risky even if exposures stay the same. Dieldrin is more lethal to the freshwater darter when temperatures are higher. Atrazine is more lethal to catfish at higher temperatures. Crabs exposed to methyl parathion are more likely to die as temperature increases.[19]

Many toxics are diluted enough so that they don't typically kill healthy fish. But these low doses may still have what biologists call sublethal effects. Low levels of toxics may affect vision or the ability to inflate swim bladders, and both may affect long-term ability to thrive. Thermal tolerance is of particular concern. Our big cold-water fish evolved to tolerate a certain amount of warmer temperatures for short periods—a trait called thermal tolerance. Climate change and endocrine-disrupting chemicals may magnify each other's effects. Researchers in Australia found that sublethal concentrations of two pesticides can significantly reduce the tolerance of some freshwater fish to increasing water temperatures. Similarly, increasing water temperatures may make it more difficult for fish to tolerate low doses of chemicals.[20]

Low-dose exposure to endocrine disruptors alters animal behavior, making affected individuals more susceptible to climate stressors that would not have harmed them otherwise. Higher blood levels of PCBs, DDT, and similar contaminants are associated with fewer viable offspring and a decrease in overwinter survival of adult birds. Many contaminants affect thyroid hormones, which in turn influence neurodevelopment and behavior, learning ability, foraging ability, hunting and survival skills, and ability to acclimate to increasing temperatures.[21]

Climate change is remobilizing contaminants that had been hidden away for decades. In the 1950s, the botany professor George Woodwell realized that the boreal forests he was studying in northern Maine had been heavily sprayed with DDT. Woodwell grew concerned, and his investigations showed that only half the DDT sprayed from the planes actually landed in the forests below. The rest seemed to vanish, and Woodwell set out to figure where it went. He learned that the DDT solution dried into tiny crystals that could be easily dispersed on air currents and eventually deposited tens of thousands of miles away. The DDT residues, Woodwell learned, were appearing not just in boreal lakes but also in the tissues of seals as far away as Antarctica.[22]

Some of the DDT and other toxics mobilized in the postwar era landed on the snows of Antarctica and the Arctic, where the crystals froze into the ice sheets and were immobilized, unable to cause harm. Until recently, that is. Global warming is now releasing those legacies of history back into the flesh of polar wildlife and from there into people. The scientist Heidi N. Geisz and her colleagues estimate that up to 2.0–8.8 pounds of DDT are released into coastal waters annually along the Western Antarctic Ice Sheet from glacial meltwater—a discovery with unnerving consequences for ecosystem health.[23]

Transport of toxics happens in the air, and it also happens on the surface of soils, roads, and buildings. Summer drought allows local pollutants to build up on surfaces. Then when an intense storm hits, heavy rains flush these toxics into Lake Superior. Those same heavy rains erode the clay soils along the south shore, leading to higher sediment and toxic loads. Many animals can adapt to small, frequent releases of pollutants, whereas sporadic pulses of high concentrations of toxics may cause more harm.

Research from extreme storm events suggests the toxic effects on fisheries may be intense. In one Seattle study, toxins in storm runoff killed Coho salmon before spawning. In another study, concentrations of all six contaminants studied increased as storms intensified. After Hurricane Katrina, levels of aldrin and other semi-volatile organic pollutants in soils increased, exceeding EPA Superfund levels. Other research found that pesticides washed off into water bodies during extreme rainfall events. These studies suggest that we should expect more storm-related contamination of Lake Superior.[24]

As waters warm, we will see more algal blooms — once very rare in Lake Superior except in the warmest shallow bays. In the summer of 2012, after an intense June storm wreaked havoc in Duluth, a plume of sediment could be traced spreading eastward into the lake over the next month. As it reached our little bay on the south shore in early July, and as temperatures soared, our clean, icy water suddenly warmed. One day I jumped off my sandstone cliff to cool off, and I smelled something grassy and pungent. Over the next week, for the first time in any local's memory, our pristine Siskiwit Bay became covered with a stinking bloom of algae. When those algal blooms decompose, they deplete oxygen, killing fish.[25] Children swimming at the beaches in the Apostle Islands National Lakeshore reported swimmer's itch, a malady of warm southern waters that support the snails that in turn support the parasites that create the itch.

Adapting to Change

What can we do about climate change and contaminants? At global levels, we may be able to mitigate some of the intensity of climate change by reducing greenhouse gases through international agreements and monetary tools such as carbon taxes and cap-and-trade. But at local levels, we need to adapt to changing climates, doing what we can to increase resiliency to the unexpected and to contain contaminants.

One key insight can help us: contaminants, even if they're persistent, bioaccumulative toxics, are not static, unchanging entities independent of biological and geological processes. They become part of earth's systems, and while they can sometimes poison those systems, biological systems can also capture them, rendering them less harmful. Certain species, for example, quickly evolve tolerance to many synthetic contaminants. Freshwater snails exposed to cadmium for only three generations evolve resistance to the cadmium. But by the fifth generation, they also show decreased tolerance to temperature stress. Similarly, Atlantic tomcod in the Hudson River evolve resistance to PCBs after years of exposure. But with the evolution of PCB resistance also comes heightened sensitivity to thermal stress and infection.[26]

Forests in the north are able to pump organochlorines and other contaminants out of the atmosphere, moving them into foliage and from there to a long-term reservoir in forest soil. This has sequestered a portion of our post–World War II toxic burden, rendering it relatively harmless. But any change that enhances the metabolism of soil organic carbon "will force POPs [persistent organic pollutants] to redistribute . . . and could result in the release of organochlorine contaminants archived in the soil's active layer during the past 50 years."[27] That would be bad news. We can prevent it if we realize that forests have captured and immobilized some industrial waste. Protecting and restoring forests will protect ourselves as well.

Planting more forests and protecting the ones we have will increase our resiliency to climate change in the Lake Superior basin. The specifics vary with location, so we need to plant forests intelligently, focusing on conifers, for example, along the north shore of Lake Superior, rather than the birch that locals love, which unfortunately are unlikely to thrive in the new climate. To reduce sediments and toxics pouring into the lake after the increasingly common heavy storms, planting forests in the upper portions of the watersheds makes a great deal of sense, particularly reforesting upper pastures growing on clay soils. County governments across the north are partnering with extension agents, tribes, and local land owners to accomplish this—a small step with the potential for big benefits.

The tribes and First Nations in the basin are already working to create a more resilient future. Wild rice is particularly important to the Anishinaabe, but it is also vulnerable to heavy spring floods and autumn drought—two increasingly frequent events predicted in most climate models for Lake Superior. Research by my graduate students, working with historical wild rice abundance data provided by the 1854 Treaty Authority, shows that heavy spring snows over the past two decades have been associated with lower wild rice abundance. But in watersheds with increasing forest cover, those associations were weaker. This research suggests that we might increase wild rice's resiliency to climate change by protecting and restoring forest cover above the wild rice lakes.

The Red Cliff Band on the Bayfield Peninsula are remediating contaminants in Lake Superior, including a program to remove some of the 1,437 barrels that the Department of Defense dumped into the lake between 1959 and 1962. Munitions waste from the Twin Cities Army Ammunition Plant had been put into fifty-five-gallon drums then dumped into Lake Superior, close to the water intake of many small communities. After a commercial fisherman snagged a barrel in 1968, the U.S. Army sent divers down to relocate some of the barrels, but most were left in place. The army and the Minnesota Pollution Control Agency insisted that the barrels posed little risk, but the tribe has chosen to pursue removal of some barrels under its authority as a sovereign nation. Initial analyses of the first 25 barrels tested found a number of toxic, persistent contaminants, including PCBs at levels higher than in nearby sediments. The barrels were rusting and some collapsed during removal, raising fears that whatever is inside those 1,437 barrels may contaminate water and sediment. After finding a chemical propellant (BLU-4) that is potentially carcinogenic (and a surprise to the Minnesota Pollution Control Agency, which had insisted nothing toxic was inside the barrels), the tribe is pursuing further steps to determine what risks the chemicals leaking from barrels might pose to Lake Superior water quality and fish consumption.[28]

Invasive species are predicted to increase, so the Bad River Band is actively controlling invasive species in the Kakagon wetlands. They are also developing a comprehensive climate change monitoring plan for the Kakagon–Bad River Sloughs, home of critical wild rice beds. Baseline data collection will allow them to better monitor changes as they happen, enabling them to actively manage habitat before changes are irreversible.[29]

On Minnesota's north shore near the Canadian border, the Grand Portage Band is promoting mixed-species forests over monocultures and planting native species to replace nonnatives. Enriching species and structural diversity will give them more options if climate variables change in surprising ways. The Fond du Lac Band near Duluth has developed an integrated resource management plan to address climate change interactions. Tribal land managers are reforesting farm fields and monitoring stream flow and water temperatures. Like other tribes, the Fond du Lac Band is reaching across government boundaries and working with local, state, and federal stakeholders on climate change issues. The Keweenaw Bay Indian Community in Michigan has hired staff to address climate change resiliency, and it is creating a climate change action plan. So far, it has created a habitat restoration project along Lake Superior to help capture contaminants washing off the shore before they get into the lake and contaminate fish.[30]

Conclusion

Many people tend to see Lake Superior as outside of history: a place so cold, remote, and pristine that little has changed for eons. It's hardly surprising that people think this. In the winter, when winds howl from the west and the temperatures drop to 20 below zero, the forests perched at the cold margins of Lake Superior can feel like the edge of the known world. Powdery snow blots out the sky when you emerge from the forest, and you can barely see through the blowing powder to the dark edge of the forest, where white pines mark the edge of abandoned fields, and black bears snooze away the winter under tip-up mounds, as they have done for thousands of years.

Yet environmental history teaches us that Lake Superior, like all the Great Lakes, is intimately connected to the rest of the globe. In the late

1950s, ornithologists discovered that DDT and other synthetic chemicals were accumulating in fish-eating birds, in bald eagles and peregrine falcons, thinning egg shells and threatening reproductive success. After years of controversy, the research eventually helped lead to global bans on DDT, PCBs, and several other persistent pollutants. Yet decades after the bans, climate change means that many of those contaminants are being released from sediments and Antarctic ice and mobilized back into global circulation. They make their way from sediments into water and then accumulate up the food chain until they are concentrated in birds. When these birds migrate, they bring toxics with them, connecting the toxic histories of far-flung places.

We are connected by the migration patterns of birds, caribou, and fish. The young loons that practice their calls on the lake bear in their bodies traces of toxins from distant shores. Migratory birds journey thousands of miles each spring to nest in the wetlands of Kakagon Sloughs — the same area at risk from mountaintop removal mines. The north is connected by currents in the atmosphere that bring pollutants from China and Africa. These currents mobilize the wastes of temperate civilizations into the fat of northern fish and mammals, and from there into the bodies of migratory birds and northern peoples. Babies in the Lake Superior basin are exposed typically to very high levels of toxins — nearly 10 percent already have toxic levels of mercury at birth. First Nations and tribal peoples are particularly vulnerable because fish are core to their cultural traditions and foodways. Flows of toxic chemicals, like the flows of energy, tie distant places together. Toxics make connections between distant places, distant times, distant peoples.

While most people assume that synthetic contaminants exist unchanging, forever poisoning the creatures that contact them, they actually may become part of biophysical systems. Like everything else, they participate in complex processes of metabolism: breaking down; creating new compounds; moving through air, water, and soil; getting safely sequestered into soil and rock and ice; or getting eroded and washed and volatilized into new mobile forms that can move across wide spaces. They become part of living ecosystems, integrating themselves into our most intimate bodily spaces —

but at the same time, ecological systems integrate them into their processes. Bacteria quickly evolve the capacity to devour synthetic contaminants and use them for energy, breaking them down in the process. Fish can quickly evolve tolerance to them, no longer becoming poisoned by their presence. Trees can capture them and sequester them, storing them safely in their roots. They are part of the systems we have become part of, as well—a radical thought to ponder.

While early efforts to control pollution in the Lake Superior basin faltered when it came to persuading industry to control effluent releases, one core idea is worth revisiting: assimilative capacity, which encompassed key understandings of natural resiliency. Early pollution models assumed that because human activities were part of larger natural processes, damage from those activities could be healed by natural processes as well. These efforts recognized the ability of ecosystems to absorb and break down pollutants, and they recognized as well the deeper idea that people are connected to the environment.

This book has traced our changing understanding of toxic mobility and its spatial relations to centers of industrial development, exploring what the archives borne within the bodies of migratory birds and fish tell us about the legacies of industrialization. All our industrial processes occur within the larger ecological and biophysical processes of earth systems, even when we have ignored those for centuries and tried to engineer them out of existence. For example, during industrialization, the assimilative capacity of streams broke down our industrial and domestic toxins, so we could keep chugging along. Atmospheric processes took away our urban and industrial toxins and transported them to the Arctic, where we thought we could ignore them, because they seemed safely stored away in permafrost, trees, soils, ice, snow. Now we are profoundly changing those processes, and the archives of our past are being lost as we remobilize the by-products of our industrial pollution back into our waters, air, and bodies. Yet restoring forests and wetlands and beaver can help restore the resiliency that will help us adapt to climate change, restore assimilative capacity, and capture and break down the persistent toxics that undermine our shared futures.

Humanistic perspectives on climate change can help us adapt. Increasingly complex climate models have been critical in helping scientists understand the unpredictable feedbacks from climate systems. Yet such models are not simple or direct representations of nature; they are filled with assumptions and uncertainties. Models that incorporate human dimensions, for example, often tend to take for granted a set of external factors, such as trade rules, intervention possibilities, and human behavior, often without acknowledging that these social factors are historically contingent rather than fixed constants.[31] When policymakers extend these models into social and cultural realms, model assumptions can lead to disastrous oversimplifications.

For example, consider the 1995 Second Assessment Report of the United Nations Intergovernmental Panel on Climate Change. This report addresses human dimensions of climate change, addressing adaptation responses, decision-making frameworks, equity, and economics. Yet adaptation considerations focus almost entirely on national-level decisions, even though the anthropologists J. McIntosh and colleagues argue that most human adaptation will come at the local or regional scale. Often, these environmental responses are based on local knowledge, which remains hidden from national governments. People survive by what the anthropologists call "networks of social relations." The large-scale technological problem-solving recommended by model outputs can disrupt such social networks, leaving people more vulnerable. In *Intimate Universality,* James Fleming notes that climate modelers have grown increasingly enamored of engineering efforts to manage and control human impacts on the climate. The history of intentional climate engineering is riddled with unintended consequences and technological hubris, making Fleming quite wary of such efforts. Smaller-scale adaptation projects, such as tree planting and wetland restoration, have less chance of badly backfiring and causing more harm than good.[32]

Historians rarely believe history can be prescriptive for policy. Yet several themes do emerge from fine-grained studies of historical responses to climate change. First, societies use their past experiences of environments—their shared environmental histories—as guides to the future.

Environmental memories help people develop ways of living in place by teaching people how to monitor land use, population levels, and economic activities. But at times of abrupt, unpredictable change, shared environmental histories may become deceptive. Historians can examine the ways people use, and misuse, environmental histories when they are trying to adapt to unpredictable change.

A second key theme is that scale matters: a society's ability to respond to environmental change varies with the scale of that change. Donald Worster argues that many modern, complex societies have learned to adapt to natural variation by concentrating "enough power and wealth at the center in order to overcome most natural vicissitudes. They learn how to create stability out of chaos by sending out money regularly to compensate for local loss."[33] While this helps minimize vulnerability to small fluctuations, such strategies may increase vulnerability to large-scale, abrupt environmental change.

One lesson from history is that people, fish, and wildlife are deeply interconnected in the Lake Superior basin. Wildlife and fisheries populations have rebounded in the basin, yet individual fish, fish-eating birds, and mammals have continued to show reproductive harm from exposure to toxics. This has posed a dilemma for some managers: should we be focusing additional recovery efforts on cleaning up toxics in sediments, or from airborne sources, when funds are limited and the issues of habitat degradation and invasive species are critical? Wildlife injuries are excellent early warning systems that alert us to potential toxic effects on people. Because we focus our ethical concerns on human individuals as well as human populations, the fact that many wildlife populations are recovering doesn't lessen the importance of toxic injury to individual animals. In other words, if we care about individuals, not just populations, we need to continue remediating persistent toxic contaminants and prevent new releases with international agreements that have real force.

What helped to clean up toxic sites in Lake Superior were the efforts of local groups as well as national and international governments. Efforts across disciplinary boundaries were critical, for when wildlife biologists began talking with toxicologists and atmospheric scientists, people began to realize the significance of global transport of contaminants. Efforts across species boundaries were equally critical, when biologists began to recognize that wildlife were sentinels for possible harm to human health.

To address the multiple stressors of climate change, habitat loss, and contaminant exposure, we need action at many levels: individual consumers working to pressure global corporations to make real changes; local and tribal governments working with the International Joint Commission to pressure Canada and the United States into concrete actions to uphold the Great Lakes Water Quality Agreement. As Rachel Carson argues in *Silent Spring,* our bodies exist in dynamic ecological relationships to ourselves and to the worlds in which we are embedded. Lake Superior's history teaches us that human histories are intimately linked to the watershed, and the quality of water determines the quality of life. When minerals are dug from the ground, when trees are cut in the forest, when flood waters are diverted, when rivers are dammed, when animals are changed from fellow creatures to livestock resources, we set into motion subtle processes of toxic transformation that have legacies far into the future. Our health emerges in relationship with larger ecological and global processes. Rich or poor, urban or rural, we are all exposed to toxic chemicals that move through our watersheds, crossing boundaries between generations, linking land and water, fish and fowl, environments and people.

Notes

Chapter One. Ecological History of the Lake Superior Basin

1. Data on relative sizes of the lakes from http://en.wikipedia.org/wiki/Lake _Superior and http://en.wikipedia.org/wiki/Lake_Erie (accessed February 3, 2017).

2. George Monro Grant, *Ocean to Ocean: Sandford Fleming's Expedition Through Canada in 1872* (Toronto: S. Low, 1877), 35. For exploration of the Great Lakes region, see John Riley, *The Once and Future Great Lakes Country: An Ecological History* (Montreal and Kingston: McGill-Queen's University Press, 2013).

3. Stephen Bocking, "Stephen Forbes, Jacob Reighard, and the Emergence of Aquatic Ecology in the Great Lakes Region," *Journal of the History of Biology* 23, no. 3 (1990): 461–498. Daniel W. Schneider, "Local Knowledge, Environmental Politics, and the Founding of Ecology in the United States: Stephen Forbes and 'The Lake as a Microcosm' (1887)," *Isis* 91, no. 4 (2000): 681–705.

4. Meg Turville-Heitz, "Superior: A Vision for the Future," Wisconsin Department of Natural Resources Report WR-346-93, 1993, 2.

5. Seth Stein et al., "Using Lake Superior Parks to Explain the Midcontinent Rift," *Park Science* 32, no. 1 (2015): 19–29. Carol A. Stein et al., "North America's Midcontinent Rift: When Rift Met LIP," *Geosphere* 11, no. 5 (2015): 1607–1616.

6. Torbjörn E. Tornqvist and Marc P. Hijma, "Links Between Early Holocene Ice-Sheet Decay, Sea-Level Rise and Abrupt Climate Change," *Nature Geoscience* 5, no. 9 (2012): 601–606. Richard B. Alley et al., "Holocene Climatic Instability: A Prominent, Widespread Event 8200 Yr Ago," *Geology* 25, no. 6 (1997): 483–486. James T. Teller, David W. Leverington, and Jason D. Mann, "Freshwater Outbursts to the Oceans from Glacial Lake Agassiz and Their Role in Climate Change During the Last Deglaciation," *Quaternary Science Reviews* 21, no. 8 (2002): 879–887.

7. Captain Thomas Jefferson Cram, *Report to Congress* (Washington D.C.: U.S. Government Printing Office, 1840); Matthew M. Thomas and Janet M. Silbernagel, "The Evolution of a Maple Sugaring Landscape on Lake Superior's Grand Island," *Michigan Academician* (2003): 135–158. For historical overviews of Lake Superior, see Norman K. Risjord, *Shining Big Sea Water:*

The Story of Lake Superior (St. Paul: Minnesota Historical Society, 2008); Patty Loew, *Indian Nations of Wisconsin: Histories of Endurance and Renewal,* 2nd ed. (Madison: Wisconsin Historical Society Press, 2013); Riley, *Great Lakes Country.*

8. Margaret Beattie Bogue, *Fishing the Great Lakes: An Environmental History, 1783–1933* (Madison: University of Wisconsin Press, 2000), 8. Charles Cleland, *Rites of Conquest: The History and Culture of Michigan's Native Americans* (Ann Arbor: University of Michigan Press, 1992). Charles Cleland, "Indians in a Changing Environment," in *The Great Lakes Forest: An Environmental and Social History,* ed. Susan Flader (Minneapolis: University of Minnesota Press, 1983).

9. Pierre Esprit Radisson, "Caesars in the Wilderness 1658," in *Up Country: Voices from the Midwestern Wilderness,* ed. William Joseph Seno (Madison: Round River, 1985), 9.

10. Alice Outwater, *Water: A Natural History* (New York: Basic Books, 1996).

11. Nicholas James Reo and Angela K. Parker, "Re-thinking Colonialism to Prepare for the Impacts of Rapid Environmental Change," *Climatic Change* 120, no. 3 (2013): doi:10.1007/s10584-013-0783-7.

12. Cleland, *Rites of Conquest.* Michael A McDonnell, *Masters of Empire: Great Lakes Indians and the Making of America* (New York: Hill and Wang, 2015). Richard White, *The Middle Ground: Indians, Empires, and Republics in the Great Lakes Region, 1650–1815* (New York: Cambridge University Press, 2011).

13. Louis Agassiz and James Elliot Cabot, *Lake Superior: Its Physical Character, Vegetation, and Animals, Compared with Those of Other and Similar Regions* (New York: Gould, 1850).

14. Ibid., 51, 104, 105, 58.

15. Ibid., 50, 96. On page 58, the authors note, "Farther on, the hills were burnt over for a great distance, showing rounded summits of white scorched rock, the lichens and earth mostly washed off from them, but the blackened tree-stems still upright." Near the Pic River: "The pitch-pine woods behind the post had been burnt over, and the trees, though yet standing, were mostly dead, affording food for myriads of wood-beetles (*Monohamus scutellaris*), whose creaking resounded on all sides. These in their turn were fed upon by the Canada jays, and by two rare species of woodpeckers," 72.

16. Ibid., 61, 74.

17. Ibid., 64.

18. Ibid., 65. On the fur trade turning to fishing, see Bogue, *Fishing the Great Lakes.*

19. William Cronon, "Kennecott Journey," in *Under an Open Sky: Rethinking America's Western Past,* ed. George Miles and Jay Gitlin (New York: Norton, 1993).

20. Larry Lankton, *Cradle to Grave: Life, Work, and Death at the Lake Superior Copper Mines* (New York: Oxford University Press, 1993). Larry Lankton, *Hollowed Ground: Copper Mining and Community Building on Lake Superior, 1840s–1990s* (Detroit: Wayne State University Press, 2010).

21. Harlan Hatcher, *A Century of Iron and Men* (New York: Bobbs-Merrill, 1950), 102. John Thistle and Nancy Langston, "Entangled Histories: Iron Ore Mining in Canada and the United States," *Extractive Industries and Society* 3, no. 2 (2016): doi:10.1016/j.exis.2015.06.003. Terry S. Reynolds and Virginia P. Dawson, *Iron Will: Cleveland-Cliffs and the Mining of Iron Ore, 1847–2006* (Detroit: Wayne State University Press, 2011).

22. Filbert Roth and Bernhard Eduard Fernow, *Forestry Conditions and Interests of Wisconsin* (Washington, D.C.: Government Printing Office, 1898). Turville-Heitz, *Superior Vision.*

23. Turville-Heitz, *Superior Vision,* 49.

24. Ibid.

25. Vernon Carstensen, *Farms or Forests: Evolution of a State Land Policy for Northern Wisconsin, 1850–1932* (Madison: College of Agriculture, University of Wisconsin, 1958).

26. Robert Gough, *Farming the Cutover: A Social History of Northern Wisconsin* (Lawrence: University Press of Kansas, 1997).

27. Ibid.

28. Flader, *Great Lakes Forest.* Carstensen, *Farms or Forests.* James Kates, *Planning a Wilderness: Regenerating the Great Lakes Cutover Region* (Minneapolis: University of Minnesota Press, 2001). Gough, *Farming the Cutover.*

29. Faith A. Fitzpatrick, James C. Knox, and Heather E. Whitman, *Effects of Historical Land-Cover Changes on Flooding and Sedimentation, North Fish Creek, Wisconsin* (Washington, D.C.: U.S. Department of the Interior, U.S. Geological Survey, 1999). Michelle Steen-Adams, Nancy Langston, and David Mladenoff, "Logging the Great Lakes Indian Reservations: The Case of the Bad River Band of Ojibwe," *American Indian Culture and Research Journal* 34, no. 1 (2010): 41–66. Michelle Steen-Adams, Nancy Langston, and David J Mladenoff, "White Pine in the Northern Forests: An Ecological and Management History of White Pine on the Bad River Reservation of Wisconsin," *Environmental History* 12, no. 3 (2007): 614–648. Michelle Steen-Adams et al., "Influence of Biophysical Factors and Differences in Ojibwe Reservation Versus Euro-American Social Histories on Forest

Landscape Change in Northern Wisconsin, USA," *Landscape Ecology* 26, no. 8 (2011): 1165–1178. Michelle Steen-Adams et al., "Historical Framework to Explain Long-Term Coupled Human and Natural System Feedbacks: Application to a Multiple-Ownership Forest Landscape in the Northern Great Lakes Region, USA," *Ecology and Society* 20, no. 1 (2015): 28.

30. Farrell M. Boyce, "Lake," in *Canadian Encyclopedia,* published December 14, 2006, www.thecanadianencyclopedia.com/articles/lake, accessed July 11, 2012.

31. "Trophic Status," Lake Access website, www.lakeaccess.org/ecology/lake ecologyprim15.html, accessed July 11, 2012.

32. Thomas A. Edsall and Murray N. Charlton, "Nearshore Waters of the Great Lakes: State of the Lakes Ecosystem Conference 1996 Background Paper," archived at http://publications.gc.ca/collections/Collection/En40-11-35-1 -1997E.pdf, accessed August 8, 2016.

33. Lake Superior Binational Program, "The Aquatic Environment," chap. 6 in *Lake Superior Lakewide Management Plan 2006* (Washington, D.C.: U.S. Environmental Protection Agency, 2006).

34. Nancy Langston, "Resiliency and Collapse: Lake Trout, Sea Lamprey, and Fisheries Management in Lake Superior," 239–262 in Lynne Heasley and Daniel Macfarlane, eds., *Border Flows: A Century of the Canadian-American Water Relationship* (Calgary: University of Calgary Press).

Chapter Two. Industrializing the Forests, 1870s to 1930s

1. J. P. Bertrand, *Timber Wolves: Greed and Corruption in Northwestern Ontario's Timber Industry, 1875–1960* (Thunder Bay, Ont.: Thunder Bay Historical Museum, 1997), 14, 74.

2. Mark Kuhlberg, "'Eyes Wide Open': E. W. Backus and the Pitfalls of Investing in Ontario's Pulp and Paper Industry, 1902–1932," *Journal of the Canadian Historical Association* 16, no. 1 (2005): 201–233.

3. The terminology is confusing for standing or logged trees that have not been processed into pulp or milled into boards. In the United States and usually in Canada, such trees are called "timber," whereas processed wood is called "lumber." But the rest of the English-speaking world uses "timber" to refer to processed wood. To avoid confusion, I will use "cut trees" or "standing trees" rather than the American term "timber" to refer to trees growing in the forest or trees that have been harvested but not yet processed. "Dimensional lumber" is wood that has been cut to standardized width and depth, typically for use in framing buildings.

4. John G. Burke, "Wood Pulp, Water Pollution, and Advertising," *Technology and Culture* 20, no. 1 (1979): 175–195.

5. Thanks to the anonymous manuscript reviewer who suggested this clarification.

6. Liza Piper, "Parasites from 'Alien Shores': The Decline of Canada's Freshwater Fishing Industry," *Canadian Historical Review* 91, no. 1 (2010): 87–114. Kuhlberg, "'Eyes Wide Open.'" Bertrand, *Timber Wolves,* 64, 66–7. For analysis of northwest Ontario's response to its hinterland status, see: G. R. Weller, "Hinterland Politics: The Case of Northwestern Ontario," *Canadian Journal of Political Science / Revue Canadienne de Science Politique* 10, no. 4 (1977): 727–754.

7. Both quotes are from Kuhlberg, "'Eyes Wide Open,'" 202–203. Kuhlberg is referring to H. V. Nelles, *The Politics of Development: Forests, Mines and Hydro-Electric Power in Ontario, 1849–1941* (Montreal and Kingston: McGill-Queen's University Press, 2005); and Peter Oliver, *G. Howard Ferguson: Ontario Tory* (Toronto: University of Toronto Press, 1977).

8. Bertrand, *Timber Wolves,* 49–50.

9. Ibid., 347, 77.

10. Ibid., 5, 15.

11. The first sulfite process pulp mill was built in Sweden in 1874. By the 1940s, the kraft process dominated paper production, and by the 2000s, it made up more than 90 percent of chemical processes. Burke, "Wood Pulp," 181–182.

12. Arn M. Keeling, "The Effluent Society: Water Pollution and Environmental Politics in British Columbia, 1889–1980" (PhD diss., University of British Columbia, 2004), 208–209. Arn Keeling, "Sink or Swim: Water Pollution and Environmental Politics in Vancouver, 1889–1975," *BC Studies: The British Columbian Quarterly* 142 no. 3 (2004): 69–101.

13. Burke, "Wood Pulp," 181.

14. Pratima Bajpai, "Generation of Waste in Pulp and Paper Mills," in *Management of Pulp and Paper Mill Waste* (Berlin: Springer International Publishing, 2015).

15. Susan L. Schantz, "Editorial: Lakes in Crisis," *Environmental Health Perspectives* 113, no. 3 (2005): A148; William D. Solecki, "Paternalism, Pollution and Protest in a Company Town," *Political Geography* 15, no. 1 (1996): 5–20; Jouni Paavola, "Water Quality as Property: Industrial Water Pollution and Common Law in the Nineteenth Century United States," *Environment and History* (2002): 295–318.

16. Samuel Wilmot, *Report on Fish-Breeding Operations in the Dominion of Canada 1890* (Ottawa: F. A. Acland, 1891); quote from Margaret Beattie Bogue,

Fishing the Great Lakes: An Environmental History, 1783–1933 (Madison: University of Wisconsin Press, 2000), 125.

17. Bogue, *Fishing the Great Lakes,* 124.
18. Burke, "Wood Pulp," 182.
19. Craig E. Colten, "Creating a Toxic Landscape: Chemical Waste Disposal Policy and Practice, 1900–1960," *Environmental History Review* 18 no. 1 (1994): 85–116.
20. Keeling, "Effluent Society," 4.
21. Roy Henry Piovesana, *Paper and People: An Illustrated History of Great Lakes Paper and Its Successors* (Thunder Bay, Ont.: Thunder Bay Historical Museum Society, 1999). Kathleen M. Wilburn and Ralph Wilburn, "Achieving Social License to Operate Using Stakeholder Theory," *Journal of International Business Ethics* 4, no. 2 (2011): 3.
22. Quote from Brent Scollie, "Pulp and Paper References from Thunder Bay Newspapers," August 1919, in Roy Piovesana, uncatalogued papers, Great Lakes Paper Company Collection, Thunder Bay Historical Museum Archives.
23. Scollie, "Pulp and Paper References," 2. Survey of the Timber Limits of the Great Lakes Paper Company, 1933, Appendix A in typescript "Early History of Great Lakes Paper Company," in Roy Piovesana, uncatalogued papers, Great Lakes Paper Company Collection, Thunder Bay Historical Museum Archives. Scollie, "Pulp and Paper References," 1.
24. Notes regarding paper mill, August 1919, and notes regarding mill, November 21, 1919, both in Correspondence Regarding Proposed Construction of a Great Lakes Pulp and Paper Company at Bare Point, 1919, box G3, folder 6, Great Lakes Paper Company Collection, Thunder Bay Historical Museum Archives.
25. "Pulp and Paper Mills at Port Arthur and Fort William," *Pulp and Paper Magazine of Canada,* January 4, 1923, 3.
26. Piovesana, *Paper and People,* 10. Mark Kuhlberg, *In the Power of the Government: The Rise and Fall of Newsprint in Ontario, 1894–1932* (Toronto: University of Toronto Press, 2015).
27. David E. Nye, *American Technological Sublime* (Cambridge: MIT Press, 1996); "Pulp and Paper Mills at Port Arthur and Fort William," 1, 4.
28. "Pulp and Paper Mills at Port Arthur and Fort William," 4.
29. "Survey of the Timber Limits," 2, 3.
30. By 1933, Great Lakes Paper controlled 9,905 square miles of timber limits, with an estimated 25,500,000 cords of pulp wood growing on it, ibid., 3.
31. Ibid., 9.

32. Ibid., 14.

33. Ibid., 16, 18, 19.

34. Piovesana, *Paper and People,* 13. Scollie, "Pulp and Paper References from Thunder Bay Newspapers," 6, 8.

35. E. A. Forsey, "The Pulp and Paper Industry," *Canadian Journal of Economics and Political Science* 1, no. 3 (1935): 501–509.

36. S. E. Peet and J. C. Day, "The Long Lake Diversion: An Environmental Evaluation," *Canadian Water Resources Journal* 5, no. 3 (1980): 34–48. "Survey of the Timber Limits," 19–21. Peter Annin, *The Great Lakes Water Wars* (Washington, D.C.: Island Press, 2009).

37. "Influences on Great Lakes Water Levels," Watershed Council, Petoskey, Michigan, at http://www.watershedcouncil.org/influences-on-great-lakes -water-levels.html (accessed February 3, 2017); Peet and Day, "Long Lake Diversion," 35. Jean Boultbee, *Pic, Pulp, and Paper: A History of the Marathon District* (Geraldton, Ont.: Town of Geraldton, 1981), 348.

38. Peet and Day, "Long Lake Diversion," 35–36.

39. Ibid., 35, 36, 44, 45.

40. Annin, *Great Lakes Water Wars.* Matthew Evenden, *Allied Power: Mobilizing Hydro-Electricity During Canada's Second World War* (Toronto: University of Toronto Press, 2015). Peet and Day, "Long Lake Diversion," 35, 39, 42.

41. Terence Kehoe, *Cleaning Up the Great Lakes: From Cooperation to Confrontation* (DeKalb: Northern Illinois University Press, 1997), 7.

42. Martin V. Melosi, "Hazardous Waste and Environmental Liability: An Historical Perspective," *Houston Law Review* 25 (1988): 741. Martin V. Melosi, *The Sanitary City: Environmental Services in Urban America from Colonial Times to the Present* (Pittsburgh: University of Pittsburgh Press, 2008).

43. Colten, "Creating a Toxic Landscape." P. W. Claassen, "Are We Abusing Our Water Resources?" *Scientific American* 124 (1921): 90.

44. George H. Ferguson, "Lake Superior," *American Journal of Public Health* 17, no. 6 (1927): 601–604. Burke, "Wood Pulp," 184.

45. Paul Wozniak, "They Thought We Were Dreamers: Early Anti-Pollution Efforts on the Lower Fox and East Rivers of Northeast Wisconsin, 1927–1949," *Transactions of the Wisconsin Academy of Sciences, Arts and Letters* 84 (1996): 161–175.

46. Park Falls Pollution Case, Flambeau River Mill, 1925, series 1243, box 8, shelf 3/43/F1–3 G4–5, Records of the Committee on Water Pollution, Wisconsin Historical Society Archives.

47. Ibid.

48. Investigation on motion of the commission of the operation of dams in and

across the Flambeau River at or near Park Falls and owned by the Flambeau Paper Company, 1925, series 1243, box 8, shelf 3/43/F1–3 G4–5, Records of the Committee on Water Pollution, Wisconsin Historical Society
Archives, 1, 2.

49. Ibid., 2.

50. Ibid., 3.

51. Ibid.

52. Ibid., 4.

53. Park Falls Pollution Case, Flambeau River Mill, 1925. John D. Rue, Forest
Products Laboratory 1925 report on pulp and paper mill waste in relation to
the pollution of streams. Presented as testimony at the Park Falls stream pol-
lution case before the Wisconsin Railroad Commission—Mr. A. Kanneberg,
October 1, 1925, series 1243, box 11, Records of the Committee on Water
Pollution, Wisconsin Historical Society Archives, 5, 7.

54. Ibid., 1.

55. L. F. Warrick and K. M. Watson, "Pollutional Surveys in Wisconsin Paper
Mills," *Paper Trade Journal* 90, no. 24 (June 12, 1930): 64–66.

56. Wozniak, "They Thought We Were Dreamers," 164–165.

57. Warrick and Watson, "Pollutional Surveys," 66. John D. Rue, Association
of Paper Manufacturers, to the Wisconsin Committee on Water Pollution,
August 29, 1929, series 1243, box 9, folder 1, Pulp and Paper Mill Surveys
1933–1934, Records of the Committee on Water Pollution, Wisconsin Historical Society Archives, 3.

58. J. M. Holderby, "Studies of Pulp and Paper Mill Wastes," *Paper Trade Journal* 61 (1933): 17–19.

59. J. M. Holderby and L. F. Warrick, "Pollutional Waste Survey of Wisconsin
Pulp and Paper Mills," *Paper Trade Journal* 64 (1935): 19–21. Pulp and Paper Advisory Committee meeting minutes, December 12, 1934, series 1243,
box 8, shelf 3/43/F1–3 G4–5, Records of the Committee on Water Pollution,
Wisconsin Historical Society Archives, 2a.

60. Wozniak, "They Thought We Were Dreamers," 166.

61. Pulp and Paper Advisory Committee meeting minutes, February 27, 1941, series 1243, box 8, shelf 3/43/F1–3 G4–5, Records of the Committee on Water
Pollution, Wisconsin Historical Society Archives, 1–2.

62. General Report on Pulp and Paper Mill Waste Surveys by the Wisconsin
State Board of Health with the Cooperation of the Wisconsin Pulp and Pa-
per Industry, 1944, series 1243, box 8, shelf 3/43/F1–3 G4–5, Records of the
Committee on Water Pollution, Wisconsin Historical Society Archives, 3.

Pulp and Paper Advisory Committee meeting minutes, February 27, 1941 and January 7, 1942.

63. Pulp and Paper Advisory Committee meeting minutes, January 7, 1942 and September 25, 1942. At this September 25, 1942, meeting, Warrick asked the industry members for information on mill wastes, stating that "it would be most desirable that at least some preliminary information on components of the kraft wastes be obtained" so they could use it in their fish tests. The industry representatives refused to provide this information.

64. Pulp and Paper Advisory Committee meeting minutes, February 2, 1945, ibid. Averill Wiley, Review of Fish Kill Study in Green Bay, 1944, Sulphite Pulp Mill's Committee on Waste Disposal, series 1243, box 11, folder 2, Records of the Committee on Water Pollution, Wisconsin Historical Society Archives.

65. Sulphite Pulp Manufacturer's Research League Report for 1962, series 1243, box 9, folder 1, Records of the Committee on Water Pollution, Wisconsin Historical Society Archives.

66. Holderby worked for the State Board of Health as assistant sanitary engineer during the 1930s. By 1941, Holderby was with the Institute of Paper Chemistry. Pulp and Paper Advisory Committee meeting minutes, November 23 and December 12, 1934, series 1243, box 8, shelf 3/43/F1 3 G4–5, Records of the Committee on Water Pollution, Wisconsin Historical Society Archives.

67. Keeling, "Effluent Society," 229.

68. Craig E. Colten and Peter N. Skinner, *The Road to Love Canal: Managing Industrial Waste Before EPA* (Austin: University of Texas Press, 1996), unpaged Kindle version. Kathryn Harrison, "The Regulator's Dilemma: Regulation of Pulp Mill Effluents in the Canadian Federal State," *Canadian Journal of Political Science / Revue Canadienne de Science Politique* 29, no. 3 (1996): 469–496. Wozniak, "They Thought We Were Dreamers," 172.

Chapter Three. The Postwar Pollution Boom

1. Craig E. Colten, "Creating a Toxic Landscape: Chemical Waste Disposal Policy and Practice, 1900–1960," *Environmental History Review* (1994): 85–116. Martin V. Melosi, *The Sanitary City: Environmental Services in Urban America from Colonial Times to the Present* (Pittsburgh: University of Pittsburgh Press, 2008).

2. For a detailed analysis of postwar recreation and environmental protection, see Paul Sutter, *Driven Wild: How the Fight Against Automobiles Launched*

the Modern Wilderness Movement (Seattle: University of Washington Press, 2009).

3. Report on Laboratory Study on the Toxicity to Fish of Four Algacides, September 6, 1949, series 1243, box 4, shelf 3/43/F1–3 G4–5, Records of the Committee on Water Pollution, Wisconsin Historical Society Archives, 17. J. B. Lackey and C. N. Sawyer, "Plankton Productivity of Certain South-Eastern Wisconsin Lakes as Related to Fertilization: I. Surveys," *Sewage Works Journal* (1945): 573–585. Aquatic Nuisance minutes, 1938–1944, July 21, 1938, 2, series 1243, box 4 folder 7, Records of the Committee on Water Pollution, Wisconsin Historical Society Archives. Report on Chemical Treatment of Lakes and Streams, April 9, 1942, 1, series 1243, box 4, shelf 3/43/F1–3 G4–5, Records of the Committee on Water Pollution, Wisconsin Historical Society Archives.

4. Report on Chemical Treatment of Lakes and Streams, April 9, 1942, 1, 14.

5. Aquatic Nuisance minutes, March 27, 1945, 7, 10, 14, 22.

6. Report on Chemical Treatment of Lakes and Streams, 14.

7. Committee on Chemical Treatment of Lakes and Streams, minutes, March 27, 1945, 1–2, series 1243, box 4, shelf 3/43/F1–3 G4–5, Records of the Committee on Water Pollution, Wisconsin Historical Society Archives.

8. Ibid., 44. Report on Laboratory Study on the Toxicity to Fish of Four Algacides, 7.

9. Arthur D. Hasler, "Eutrophication of Lakes by Domestic Drainage," *Ecology* 28, no. 4 (1947): 383–395.

10. Kenneth M. Mackenthun, Report on the Chemical Treatment Program, 1951, 2. Regular meeting Sub-Committee on Aquatic Nuisance Control 1964–66, March 30, 1964, 5. Both in series 1243, box 4, shelf 3/43/F1–3 G4–5, Records of the Committee on Water Pollution, Wisconsin Historical Society Archives.

11. Report on Laboratory Study on the Toxicity to Fish of Four Algacides, 7, 16.

12. Anders Halverson, *An Entirely Synthetic Fish: How Rainbow Trout Beguiled America and Overran the World* (New Haven: Yale University Press, 2010). Robert A. Hughes and G. Fred Lee, "Toxaphene Accumulation in Fish in Lakes Treated for Rough Fish Control," *Environmental Science & Technology* 7, no. 10 (1973): 934.

13. G. B. Frankforter and F. C. Frary, "The Chlor-Hydrochlorides of Pinene and Firpene," *Journal of the American Chemical Society* 28, no. 10 (1906): 1461–1467. W. L. Parker and J. H. Beacher, "Toxaphene: A Chlorinated Hydrocarbon with Insecticidal Properties," University of Delaware Agricul-

tural Experiment Station, Bulletin No. 264, 1947. E. G. Batte and R. D. Turk, "Toxicity of Some Synthetic Insecticides to Dogs," *Journal of Economic Entomology* 41, no. 1 (1948): 102–103. Hercules Powder Company, *More Cotton, More Profit with Toxaphene (Hercules Chlorinated Camphene)* (Wilmington, Del.: Hercules Powder Company, 1948).

14. Frank F. Hooper and Alfred R. Grzenda, "The Use of Toxaphene as a Fish Poison," *Transactions of the American Fisheries Society* 85, no. 1 (1957): 180–190. Jack E. Hemphill, "Toxaphene as a Fish Toxin," *Progressive Fish-Culturist* 16, no. 1 (1954): 41–42. Burton J. Kallman, Oliver B. Cope, and Richard J. Navarre, "Distribution and Detoxication of Toxaphene in Clayton Lake, New Mexico," *Transactions of the American Fisheries Society* 91, no. 1 (1962): 14–22.

15. For a detailed analysis, see Nancy Langston, *Toxic Bodies: Hormone Disruptors and the Legacy of DES* (New Haven: Yale University Press, 2010).

16. Colten, "Creating a Toxic Landscape," 94–95.

17. Alan Hall, "Interview with Barry Commoner," *Scientific American*, June 23, 1997.

18. John Bellamy Foster and Brett Clark, "Rachel Carson's Ecological Critique," *Monthly Review* 59, no. 9 (2008).

19. Theodore R. Rice, "The Accumulation and Exchange of Strontium by Marine Planktonic Algae," *Limnology and Oceanography* 1, no. 2 (1956): 123–138. J. J. Davis and R. F. Foster, "Bioaccumulation of Radioisotopes Through Aquatic Food Chains," *Ecology* 39, no. 3 (1958): 530–535. W. C. Hanson, "Accumulation of Radioisotopes from Fallout by Terrestrial Animals at Hanford, Washington," *Northwest Science (US)* 34 (1960).

20. Ralph H. Lutts, "Chemical Fallout: Rachel Carson's Silent Spring, Radioactive Fallout, and the Environmental Movement," *Environmental Review* 9, no. 3 (1985): 211–225. Rachel Carson, *Silent Spring* (New York: Houghton Mifflin, 2002).

21. Rachel Carson, "The Pollution of Our Environment," 227–245 in Linda Lear, ed., *Lost Woods: The Discovered Writing of Rachel Carson* (Boston: Beacon Press, 2011), 238.

22. Roy Piovesana, *Paper and People: An Illustrated History of Great Lakes Paper and Its Successors* (Thunder Bay, Ont.: Thunder Bay Historical Museum Society, 1999), 39. R. A. Wheatley, "An Industry of National Importance: The Origin and Development of the Great Lakes Paper Company Limited," *Pulp and Paper Magazine of Canada* (1943): 559–578.

23. Nancy Langston, "Paradise Lost: Climate Change, Boreal Forests, and Environmental History," *Environmental History* 14, no. 4 (2009): 641–650.

24. See George Woodwell, "Toxic Food Web," in Heather Newbold, *Life Stories: World-Renowned Scientists Reflect on Their Lives and on the Future of Life on Earth* (Berkeley: University of California Press, 2000), 74. For the original research, see G. M. Woodwell and F. T. Martin, "Persistence of DDT in Soils of Heavily Sprayed Forest Stands," *Science* 145, no. 3631 (1964): 481–483.

25. Nicholas C. Bolgiano, "Changes in Boreal Bird Irruptions in Eastern North America Relative to the 1970s Spruce Budworm Infestation," *American Birds* 54 (2004): 26–33.

26. Ibid.

27. Woodwell and Martin, "Persistence of DDT."

28. Robie Macdonald et al., "Contaminants in the Canadian Arctic: Five Years of Progress in Understanding Sources, Occurrence and Pathways," *Science of the Total Environment* 254, no. 2 (2000): 93–234. Robie Macdonald et al., "How Will Global Climate Change Affect Risks from Long-Range Transport of Persistent Organic Pollutants?" *Human and Ecological Risk Assessment* 9, no. 3 (2003): 643–660.

29. Macdonald et al., "Contaminants in the Canadian Arctic," 102.

30. Ibid., 96–97.

31. Ibid., 130. R. W. Macdonald, T. Harner, and J. Fyfe, "Recent Climate Change in the Arctic and Its Impact on Contaminant Pathways and Interpretation of Temporal Trend Data," *Science of the Total Environment* 342, no. 1–3 (2005): 5–86.

32. Ben K. Greenfield et al., "Predicting Mercury Levels in Yellow Perch: Use of Water Chemistry, Trophic Ecology, and Spatial Traits," *Canadian Journal of Fisheries and Aquatic Sciences* 58, no. 7 (2001): 1419–1429.

33. Macdonald et al., "Contaminants in the Canadian Arctic," 130. Macdonald, Harner, and Fyfe, "Recent Climate Change in the Arctic," 50. Richard Bindler et al., "Mining, Metallurgy and the Historical Origin of Mercury Pollution in Lakes and Watercourses in Central Sweden," *Environmental Science & Technology* 46, no. 15 (2012): 7984–7991. H. R. Friedli et al., "Initial Estimates of Mercury Emissions to the Atmosphere from Global Biomass Burning," *Environmental Science & Technology* 43, no. 10 (2009): 8092–8098. Daniel R. Engstrom, Steven J. Balogh, and Edward B. Swain, "History of Mercury Inputs to Minnesota Lakes: Influences of Watershed Disturbance and Localized Atmospheric Deposition," *Limnology and Oceanography* 52, no. 6 (2007): 2467–2483. Paul E. Drevnick et al., "Spatial and Temporal Patterns of Mercury Accumulation in Lacustrine Sediments Across the Laurentian Great Lakes Region," *Environmental Pollution* 161 (2012): 252–260. Mark E. Brigham et al., "Mercury Cycling in Stream Ecosystems. 1. Water

Column Chemistry and Transport," *Environmental Science & Technology* 43, no. 8 (2009): 2720–2725.

34. Vicki Monks, "Dioxin's Toll on Wildlife," *Synthesis/Regeneration* 7, no. 8 (1995). Ellen Griffith Spears, *Baptized in PCBs: Race, Pollution, and Justice in an All-American Town* (Chapel Hill: University of North Carolina Press, 2014). Barry Commoner, "The Political History of Dioxin," *Synthesis/Regeneration* 7, no. 8 (1995).

35. Ronald A. Hites, "Dioxins: An Overview and History," *Environmental Science & Technology* 45, no. 1 (2010): 16–20.

36. Monks, "Dioxin's Toll on Wildlife." Michael Gilbertson et al., "Great Lakes Embryo Mortality, Edema, and Deformities Syndrome (GLEMEDS) in Colonial Fish-Eating Birds: Similarity to Chick-Edema Disease," *Journal of Toxicology and Environmental Health, Part A Current Issues* 33, no. 4 (1991): 455–520.

37. Hites, "Dioxins." Carol Van Strum, "EPA Reinvents the Wheel on Reproductive Effects of Dioxin," *Synthesis/Regeneration* 7, no. 8 (1995).

38. Carol Van Strum and Paul Merrell, *No Margin of Safety: A Preliminary Report on Dioxin Pollution and the Need for Emergency Action in the Pulp and Paper Industry* (Toronto: Greenpeace Great Lakes, 1987), section IV, 5–7.

39. Commoner, "Political History of Dioxin."

40. Van Strum and Merrell, *No Margin of Safety,* section V, 6–8.

41. Ibid., section III, 13; Philip Shabecoff, "Government Says Dioxin from Paper Mills Poses No Major Danger," *New York Times,* May 1, 1990.

42. Grand Council Treaty #3, Interrogatories and replies to interrogatories: witness statement I to VI, 1987, In The Matter of a hearing before the Environmental Assessment Board regarding a Class Environmental Assessment for Timber Management on Crown Lands in Ontario, in Regional Collections, Lakehead University Archives, Thunder Bay, Ont. Albert Ernest Jenks, *The Wild Rice Gatherers of the Upper Lakes: A Study in American Primitive Economics* (Washington, D.C.: U.S. Government Printing Office, 1901). Thomas Vennum, *Wild Rice and the Ojibway People* (St. Paul: Minnesota Historical Society Press, 1988), 266–267.

43. Grand Council Treaty #3, 2, 3, 8.

44. On sturgeon, see Vennum, *Wild Rice and the Ojibway People,* 287. On dams, see Complaints, NAC, RG 10, V. 7842, F. 30124–4, Chief and Councillors of Little Fork Reserve to R. L. Borden, premier of Canada, June 2, 1913, cited in Grand Council Treaty #3, 11. On wild rice: ibid., 11.

45. Beverley Chalmers and Shi Wu Wen, "Perinatal Care in Canada," *BMC Women's Health* 4, no. 1 (2004): 1. Jill Torrie et al., "The Evolution of

Health Status and Health Determinants in the Cree Region (Eeyou Istchee):
Eastmain-1-A Powerhouse and Rupert Diversion Sectoral Report," *Report
on the Health Status of the Population,* ser. 4, no. 3: Cree Board of Health and
Social Services (2005).

46. Matthew P. Longnecker et al., "Association Between Maternal Serum
    Concentration of the DDT Metabolite DDE and Preterm and Small-for-
    Gestational-Age Babies at Birth," *Lancet* 358, no. 9276 (2001): 110–114.

47. John Rudd, Reed Harris, and Patricia Sellers, "Advice on Mercury Reme-
    diation Options for the Wabigoon-English River System Final Report,"
    prepared for Asubpeeschoseewagong Netum Anishinabek (March 21, 2016),
    available at http://freegrassy.net/wp-content/uploads/2016/05/Wabigoon
    -English-River-System-Advice-on-Mercury-Remediation-Final-March-21a.
    pdf. Maya Basdeo and Lalita Bharadwaj, "Beyond Physical: Social Dimen-
    sions of the Water Crisis on Canada's First Nations and Considerations for
    Governance," *Indigenous Policy Journal* 23, no. 4 (2013). Patricia A. D'Itri,
    "The Distribution of Mercury," in P. A. Krenkel, ed., *Heavy Metals in the
    Aquatic Environment: Proceedings of the International Conference Held in
    Nashville, Tennessee, December 1973* (Elmsford, N.Y.: Pergamon Press, 1975).
    Ryan Duplassie, "Idle No More: Indigenous People's Coordinated Reac-
    tion to the Twin Forces of Colonialism and Neo-Colonialism in Canada,"
    in Sophie Croisy, ed., *Globalization and "Minority" Cultures* (Leiden: Brill,
    2014). Michael Egan, "Chronicling Quicksilver's Anthropogenic Cycle,"
    Global Environment 7, no. 1 (2014): 10–37. Natalia Ilyniak, "Colonialism
    and Relocation: An Exploration of Genocide and the Relocation of Animist
    Aboriginal Groups in Canada," *Journal of Religion and Culture: Conference
    Proceedings,* 17th Annual Graduate Interdisciplinary Conference, Concordia
    University, Montreal (March 2012): 75–83, available at https://ojs.concordia.
    ca/index.php/jrc/article/viewFile/271/180. Natalia Ilyniak, "Mercury Poi-
    soning in Grassy Narrows: Environmental Injustice, Colonialism, and Capi-
    talist Expansion in Canada," *McGill Sociological Review* 4 (2014): 43–66.
    Anastasia M. Shkilnyk, *A Poison Stronger Than Love: The Destruction of
    an Ojibwa Community* (New Haven: Yale University Press, 1985). Shigeru
    Takaoka et al., "Signs and Symptoms of Methylmercury Contamination in a
    First Nations Community in Northwestern Ontario, Canada," *Science of the
    Total Environment* 468 (2014): 950–957.

48. "Mercury Poisoning Effects Continue at Grassy Narrows," CBCNews Can-
    ada (June 4, 2012). G. R. Weller, "Hinterland Politics: The Case of North-
    western Ontario," *Canadian Journal of Political Science / Revue Canadienne
    de Science Politique* 10, no. 4 (1977): 727–754.

Chapter Four. Taconite and the Fight over
Reserve Mining Company

1. "The Iron Ore Dilemma," *Fortune,* December 1945, 129–131. John Thistle and Nancy Langston, "Entangled Histories: Iron Ore Mining in Canada and the United States," *Extractive Industries and Society* 3, no. 2 (2016): 269–277.

2. Thistle and Langston, "Entangled Histories." Terry Reynolds and Virginia Dawson, *Iron Will: Cleveland-Cliffs and the Mining of Iron Ore, 1847–2006* (Detroit: Wayne State University Press, 2011). W. Bruce Bowlus, *Iron Ore Transport on the Great Lakes: The Development of a Delivery System to Feed American Industry* (Jefferson, N.C.: McFarland, 2010). Robert V. Bartlett, *The Reserve Mining Controversy: Science, Technology, and Environmental Quality* (Bloomington: Indiana University Press, 1980).

3. Mesabi Taconite and the War Emergency, April 1, 1942, report by University of Minnesota Mines Experiment Station, 127.I.4.6F, box 1–3, Minnesota Attorney General, Natural Resources Division, Reserve Mining Company Files, State Archives, Minnesota Historical Society, 7. Reynolds and Dawson, *Iron Will.* Richard B. Mancke, "Iron Ore and Steel: A Case Study of the Economic Causes and Consequences of Vertical Integration," *Journal of Industrial Economics* (1972): 220–229.

4. Jeffrey T. Manuel, "Mr. Taconite: Edward W. Davis and the Promotion of Low-Grade Iron Ore, 1913–1955," *Technology and Culture* 54, no. 2 (2013): 317–345. Timothy LeCain, *Mass Destruction: The Men and Giant Mines That Wired America and Scarred the Planet* (Rutgers: Rutgers University Press, 2009). Jeffrey T. Manuel, *Taconite Dreams: The Struggle to Sustain Mining on Minnesota's Iron Range, 1915–2000* (St. Paul: University of Minnesota Press, 2015).

5. John Baeten and Nancy Langston, "Mapping Mines to Remember Waste: A Spatial History of Iron Beneficiation in the Lake Superior Basin," Poster Presentation, American Society for Environmental History Conference, Seattle, Washington: April, 2016.

6. Ibid.

7. Michael E. Berndt and William C. Brice, "The Origins of Public Concern with Taconite and Human Health: Reserve Mining and the Asbestos Case," *Regulatory Toxicolology and Pharmacology* 52, no. 1 suppl. (2008): S31–39.

8. Application of Reserve Mining Company to Water Pollution Control Commission, 1947, 117.G.17.1 MC, Minnesota Attorney General, Natural Resources Division, Reserve Mining Company Files, State Archives, Minnesota Historical Society, 3.

9. LeCain, *Mass Destruction*.

10. Manuel, "Mr. Taconite," 320. Bartlett, *Reserve Mining Controversy*, 20.

11. James C. Scott, *Seeing Like a State: How Certain Schemes to Improve the Human Condition Have Failed* (New Haven: Yale University Press, 1998). Taylor testimony in Hearing Transcripts (Department of Conservation), 117.G.17.1 MC application for permit, Minnesota Attorney General, Natural Resources Division, Reserve Mining Company Files, State Archives, Minnesota Historical Society (hereafter Hearing Transcripts), July 1947, 41. Edward H. Davis, *Pioneering with Taconite* (St. Paul: Minnesota Historical Society Press, 2004).

12. Thomas R. Huffman, "Enemies of the People: Asbestos and the Reserve Mining Trial," *Minnesota History* 59, no. 7 (2005): 292–306.

13. Taylor testimony in Hearing Transcripts, July 1947, 39.

14. Wilson's obituary in the *New York Times* noted, "Taking a firm states-rights position, Wilson refused to accept the idea that Minnesota should hand over title to the extensive state forest lands [to Quetico-Superior National Parks]. . . . He opposed the notion of a Canadian-American treaty to protect the watershed, since it would require undue federal interference in state lands." Cited in Gerald Killan, *Protected Places: A History of Ontario's Provincial Parks System* (Toronto: Dundurn, 1993), 89.

15. Application of Reserve Mining Company to Water Pollution Control Commission, 117.G.17.1 MC application for permit, 1947, Minnesota Attorney General, Natural Resources Division, Reserve Mining Company Files, State Archives, Minnesota Historical Society, 6.

16. Ibid., 3–4.

17. "Iron Ore Dilemma." Manuel, "Mr. Taconite."

18. Edward Davis testimony in Hearing Transcripts, July 1947, 48–49.

19. Quotes ibid., 37–38 and 56–57.

20. Mr. Sharp testimony in Hearing Transcripts, July 1947, 87–89.

21. Hearing Transcripts, July 1947, 107–108.

22. Ibid., 7. Mesabi Taconite and the War Emergency, April 1, 1942, report by University of Minnesota Mines Experiment Station, 127.I.4.6F, box 1–3, Minnesota Attorney General, Natural Resources Division, Reserve Mining Company Files, State Archives, Minnesota Historical Society, 18. My analysis of Reserve's naturalness debates relies on envirotech perspectives which view technological systems (such as mining) as hybrids that blur the boundaries between natural and synthetic; see LeCain, *Mass Destruction*.

23. Taylor testimony in 1947 Hearing Transcripts, 27.

24. Hearing Transcripts, July 1947, 7–8, 11. Taylor ibid., 35. Ibid., 54.

25. Hearing Transcripts, City of Duluth, 1947, 117.G.17.1 MC application for permit, 1947, Minnesota Attorney General, Natural Resources Division, Reserve Mining Company Files, State Archives, Minnesota Historical Society, 96–97.

26. John Moyle, ibid., 74–75.

27. Hearing Transcripts, July 1947, 14–15. Hearing Transcripts, City of Duluth, 1947, 130.

28. Davis, *Pioneering with Taconite*, 128.

29. Hearing Transcripts, City of Duluth, 1947, 131. 1947 Hearing Transcripts, 103. Hearing Transcripts, July 1947, 34.

30. Hearing Transcripts, City of Duluth, July 1947, 26.

31. Mark Walrod Harrington, Norman B. Conger, and R. F. De Grain, *Surface Currents of the Great Lakes: As Deduced from the Movements of Bottle Papers During the Seasons of 1892, 1893, and 1894* (Washington, D.C.: Government Printing Office, Weather Bureau, 1895). The bathtubs quote comes from *Living with the Lakes: Understanding and Adapting to the Great Lakes Water Level Changes* (Detroit: U.S. Army Corps of Engineers Detroit District and Great Lakes Commission, 2000), 19.

32. Hearing Transcripts, City of Duluth, 1947, 44–45. Thomas F. Bastow, *This Vast Pollution: United States of America v. Reserve Mining Company* (Washington, D.C.: Green Fields Books, 1986), 7.

33. Correspondence, 1947 permit hearings, 117.G.17.1 MC application for permit, Minnesota Attorney General, Natural Resources Division, Reserve Mining Company Files, State Archives, Minnesota Historical Society.

34. John Moyle testimony, 1947 Hearing Transcripts, 75. Hearing Transcripts, July 1947, 18. John Moyle, in Hearing Transcripts, City of Duluth, 1947, 60.

35. Testimony of Engineer Engells in 1947 Hearing Transcripts, 71, 72, 74. Moyle in 1947 Hearing Transcripts, 80.

36. Testimony by Mr. Tuskey in Hearing Transcripts, September 30, 1947 (vol. VIa). Bastow, *This Vast Pollution*, 6. Testimony of McGath and Jensen in Hearing Transcripts, July 1947 (vol. IV, July 22), 43.

37. Bastow, *This Vast Pollution*, 8.

38. Jensen testimony in Hearing Transcripts, July 1947 (vol. IV, July 22), 28–29, 30, 36.

39. Hearing Transcripts, City of Duluth, 1947, 140.

40. Ibid., 143.

41. Ibid., 144.

42. Ibid., 98 and 124.

43. Ibid., 116.

44. Laing, ibid., 113–116; Kaupanger and McGath quotes, ibid., 90.

45. Wilson, summary of evidence, 117.G.17.1 MC application for permit, January 1947, Minnesota Attorney General, Natural Resources Division, Reserve Mining Company Files, State Archives, Minnesota Historical Society, 6.

46. Hearing Transcripts, September 30, 1947, morning (vol. VIa), 219.

47. Commissioner Wilson to Governor Youngdahl, July 2, 1947, in Correspondence Before Hearings, 117.G.17.1 MC application for permit, 1947, Minnesota Attorney General, Natural Resources Division, Reserve Mining Company Files, State Archives, Minnesota Historical Society.

48. Senator Carr to Commissioner Wilson, July 29, 1947, ibid.

49. Commissioner Wilson to Senator Carr, July 31, 1947, ibid.

50. Davis in Hearing Transcripts, July 1947.

51. Bartlett writes: "The state of scientific knowledge about Lake Superior in 1947 was inadequate for the decision that was made. Most of the testimony about the heavy density current was based on laboratory experiments. . . . Little was known about Lake Superior, biologically or chemically—and as it turned out, there was considerable ignorance about the chemical and physical properties of the tailings as well. . . . The researchers at the University of Minnesota clearly saw what they wanted to see in their experiments on heavy density currents" (*Reserve Mining Controversy*, 28).

52. Hearing Transcripts, July 1947, 27. Hearing Transcripts, City of Duluth, 1947, 65.

53. Commissioner Wilson to Senator Carr, January 12, 1948, in Correspondence Before Hearings, 117.G.17.1 MC application for permit, 1947, Minnesota Attorney General, Natural Resources Division, Reserve Mining Company Files, State Archives, Minnesota Historical Society.

54. Ibid.

55. Ibid.

Chapter Five. Mining Pollution Debates, 1950s Through the 1970s

1. First permit revision, see Reserve Mining Company application for amendment to permit, June 27, 1956, 117.G.17.1, box I-1, Minnesota Attorney General, Natural Resources Division, Reserve Mining Company Files, State Archives, Minnesota Historical Society. Second permit revision, see Second Amendment to Permit 1960, 117.G.17.1, box I-1 Minnesota Attorney General, Natural Resources Division, Reserve Mining Company Files, State Archives, Minnesota Historical Society. Correspondence regarding tailing launder design, 1954–1966, 117.G.17.9. For the fly ash permit revision, see

117.G.17.9, Correspondence 1954–1979, Minnesota Attorney General, Natural Resources Division, Reserve Mining Company Files, State Archives, Minnesota Historical Society.

2. Luther J. Carter, "Pollution and Public Health: Taconite Case Poses Major Test," *Science* 186, no. 4158 (1974): 31–36.

3. Thomas F. Bastow, *This Vast Pollution: United States of America v. Reserve Mining Company* (Washington, D.C.: Green Fields Books, 1986), 9.

4. George A. Selke to Sidney Frellsen, 1956, 117.G.17.1, box I-1, Minnesota Attorney General, Natural Resources Division, Reserve Mining Company Files, State Archives, Minnesota Historical Society. Water Pollution Control Commission: Hearing Transcript, September 25, 1956, 117.G.17.1, box I-1, Minnesota Attorney General, Natural Resources Division, Reserve Mining Company Files, State Archives, Minnesota Historical Society.

5. Ojard testimony in Water Pollution Control Commission: Hearing Transcript, September 25, 1956, 117.G.17.1, box I-1, Minnesota Attorney General, Natural Resources Division, Reserve Mining Company Files, State Archives, Minnesota Historical Society, 107, 111, 114.

6. Lawrence Bugge testimony ibid., 19.

7. Ruble quote from Selke to Frellsen, 1956.

8. Bastow, *This Vast Pollution,* 10–11.

9. Second Amendment to Permit 1960, 3 4. Robert J. Linney testimony in Water Pollution Control Commission: Hearing Transcript, July 13, 1960, 117.G.17.1, box I-1, Minnesota Attorney General, Natural Resources Division, Reserve Mining Company Files, State Archives, Minnesota Historical Society, 14, 1–2. Asbestos terminology is confusing because "asbestos" is an industrial term that refers to a group of mineral forms composed of long, thin fibers that are flexible and can easily be separated from each other. Most types of asbestos belong to the amphibole group of minerals, which are fibrous minerals composed of parallel chains of silica needles. Asbestiform fibers (including asbestos and similar minerals) have certain structural properties that can cause lung damage. For an overview, see National Research Council (U.S.) Committee on Nonoccupational Health Risks of Asbestiform Fibers, *Asbestiform Fibers: Nonoccupational Health Risks* (Washington, D.C.: National Academies Press, 1984).

10. Barbara L. Allen, "Shifting Boundary Work: Issues and Tensions in Environmental Health Science in the Case of Grand Bois, Louisiana," *Science as Culture* 13, no. 4 (2004): 429–448.

11. Edward Davis to Mr. C. L. Kingsbury, vice president of Reserve, December 7, 1954, Correspondence regarding tailing launder design, 117.G.17.9,

Correspondence 1954–1979, Minnesota Attorney General, Natural Resources Division, Reserve Mining Company Files, State Archives, Minnesota Historical Society. Davis wrote: "Changes in the tailings launder or water intake should be made only after we see whether or not trouble develops. After we get into operation, and if trouble develops, then we can call in experts for advice as to what to do."

12. February 21, 1966, ibid.

13. Memo from Edward Schmid to J. Wm. Bryant, February 2, 1966, ibid.

14. L. O. Mayer, general superintendent of Power Production, Northern States Power Company, to Edward Schmid, director of Reserve public relations, October, 10, 1967, ibid.

15. Handwritten field notes titled "Reserve Mining Fly Ash Sample, with rankings of toxicity, and effects on biological life," September 10, 1969, ibid. For other examples of suppression of scientific findings, see Allan Brandt, *The Cigarette Century: The Rise, Fall, and Deadly Persistence of the Product That Defined America* (New York: Basic Books, 2009); Nancy Langston, *Toxic Bodies: Hormone Disruptors and the Legacy of DES* (New Haven: Yale University Press, 2010); Gerald Markowitz and David Rosner, *Deceit and Denial: The Deadly Politics of Industrial Pollution* (Berkeley: University of California Press, 2013); David Michaels, *Doubt Is Their Product: How Industry's Assault on Science Threatens Your Health* (New York: Oxford University Press, 2008); Naomi Oreskes and Erik M. Conway, *Merchants of Doubt: How a Handful of Scientists Obscured the Truth on Issues from Tobacco Smoke to Global Warming* (New York: Bloomsbury, 2011).

16. "Taconite Pellets Put Mesabi Back to Work," *Chemical Engineering News Archive* 45, no. 49 (1967): 60–64. Humphrey quoted in "Vice President Urges Teachers to Open Science to All," *Chemical Engineering News Archive* 46, no. 16 (1968): 20.

17. Robert V. Bartlett, *The Reserve Mining Controversy: Science, Technology, and Environmental Quality* (Bloomington: Indiana University Press, 1980), 43.

18. Stephen W. Stathis, *Landmark Legislation, 1774–2012: Major U.S. Acts and Treaties* (Washington, D.C.: CQ Press, 2014), 278.

19. "Better Late Than Never: The Lake Superior Story," undated, in 117.G.18.4F, box III-4, Minnesota Attorney General, Natural Resources Division, Reserve Mining Company Files, State Archives, Minnesota Historical Society, 24. A revised version of this essay was published as chapter 7 in David Zwick and Marcy Benstock, *Water Wasteland: Ralph Nader's Study Group Report on Water Pollution* (New York: Grossman Publishers, 1971).

20. Bruce Lambert, "John A. Blatnik, 80, Congressman Who Promoted Public Works Bills," *New York Times,* December 19, 1991.

21. "Better Late Than Never," 6–7.

22. Ibid., 26. Bastow, *This Vast Pollution,* 11.

23. Bastow, *This Vast Pollution,* 9.

24. "Better Late Than Never," 29. Bastow, *This Vast Pollution,* 9.

25. Freeman Holmer (assistant to Governor Knowles) to Don Mount, October 16, 1967. Response Mount to Holmer, October 18, 1967; both in Correspondence 1954–1979, 117.G.17.9, Minnesota Attorney General, Natural Resources Division, Reserve Mining Company Files, State Archives, Minnesota Historical Society.

26. Mount to Williams, January 26, 1968, in Correspondence re Stoddard Report and others, 117.G.17.9, box II-4, Minnesota Attorney General, Natural Resources Division, Reserve Mining Company Files, State Archives, Minnesota Historical Society.

27. Louis G. Williams, "Should Some Beneficial Uses of Public Waterways Be Illegitimate?," *BioScience* 18, no. 1 (1968): 36–37.

28. Swift article in *Conservation News,* newsletter of the National Wildlife Federation, vol. 32, no. 24 (December 15, 1967), in folder 23, Public Relations, Reserve Mining Company, regarding Stoddard Report Files, 117.G.17.9, box II-8, Minnesota Attorney General, Natural Resources Division, Reserve Mining Company Files, State Archives, Minnesota Historical Society.

29. Memo from E. M. Furness to Edward Schmid, Reserve, January 3, 1986, ibid.

30. Ibid.

31 Terence Kehoe, *Cleaning Up the Great Lakes: From Cooperation to Confrontation* (DeKalb: Northern Illinois University Press, 1997).

32. Bartlett, *Reserve Mining Controversy,* 33.

33. Ibid. Charles Stoddard, Special Report on Water Quality of Lake Superior in the Vicinity of Silver Bay, December 1968, 2 folders, 117.G.17.9, Minnesota Attorney General, Natural Resources Division, Reserve Mining Company Files, State Archives, Minnesota Historical Society (hereafter Stoddard Report).

34. Bill Berry, "Herb Behnke, Martin Hanson and Charles Stoddard—Our Newest Inductees," *New Leaf: Newsletter of the Wisconsin Conservation Hall of Fame* 15, no. 1 (2009): 1–3. Bastow, *This Vast Pollution,* 14.

35. Ibid., 16. W. K. Montague, Statement by Reserve Mining Company to the Corps of Engineers on the Stoddard Report process, November 12, 1968,

Correspondence 1954–1979, 117.G.17.9, Minnesota Attorney General, Natural Resources Division, Reserve Mining Company Files, State Archives, Minnesota Historical Society, 10.

36. John L. Skrypek et al., "Bottom Fauna off the Minnesota North Shore of Lake Superior as Related to Deposition of Taconite Tailing and Fish Production," State of Minnesota Department of Conservation Division of Game and Fish, in cooperation with the Minnesota Pollution Control Agency, Special Publication no. 57, October 10, 1968, 117.G.17.1, Minnesota Attorney General, Natural Resources Division, Reserve Mining Company Files, State Archives, Minnesota Historical Society. Material on Report on Water Quality and Sources of Wastes in the Lake Superior Basin, Minnesota Pollution Control Agency, April 1969, 117.G.17.5, box II-4, Minnesota Attorney General, Natural Resources Division, Reserve Mining Company Files, State Archives, Minnesota Historical Society, 5.

37. Material on Report on Water Quality and Sources of Wastes in the Lake Superior Basin, Minnesota Pollution Control Agency, April 1969, 117.G.17.5, box II-4, Minnesota Attorney General, Natural Resources Division, Reserve Mining Company Files, State Archives, Minnesota Historical Society, 1. John Badalich to Charles Stoddard, October 10, 1968, and Charles Stoddard to John Badalich, October 16, 1968, in Correspondence, Stoddard Report files, 117.G.17.9, box II-8, folder 22, Minnesota Attorney General, Natural Resources Division, Reserve Mining Company Files, State Archives, Minnesota Historical Society.

38. Mount, Lennon, and Pycha responses to Reserve comments on Stoddard Report, October 18, 1969, in Correspondence, Stoddard Report files, 117.G.17.9, box II-8, folder 22, Minnesota Attorney General, Natural Resources Division, Reserve Mining Company Files, State Archives, Minnesota Historical Society. Bastow, *This Vast Pollution*, 18.

39. Stoddard, Stoddard Report, 22, 26–27.

40. Ibid., 23.

41. Bartlett, *Reserve Mining Controversy*, 59. Fred Lee, Reserve consultant, to H. Zar of Federal Water Pollution Control Agency, June 26, 1970, in Correspondence 1954–1979, 117.G.17.9, Minnesota Attorney General, Natural Resources Division, Reserve Mining Company Files, State Archives, Minnesota Historical Society.

42. Bartlett, *Reserve Mining Controversy*, 60–61.

43. "Better Late Than Never," 8. 104 Cong. Rec., 105th Congress, 36, statement of Dean Rebuffoni. Bastow, *This Vast Pollution*, 20. Bartlett, *Reserve Mining Controversy*, 59. Charles Stoddard to Chief of U.S. Army Corps, Janu-

ary 1, 1969, in Gaylord Nelson Papers, Subseries John Heritage, MSS 1020, box 131, Wisconsin Historical Society Archives.

44. "Pollution of Lake Superior," editorial in *Washington Post,* May 14, 1969. Reserve Mining Company comments on revised Stoddard Report, April 22, 1969, 117.G.17.9, Minnesota Attorney General, Natural Resources Division, Reserve Mining Company Files, State Archives, Minnesota Historical Society. Don Wright of Reserve, memo to Ed Schmidt of Reserve, April 22, 1969, in Correspondence, Stoddard Report Files, 117.G.17.9, box II-8, folder 22, Minnesota Attorney General, Natural Resources Division, Reserve Mining Company Files, State Archives, Minnesota Historical Society.

45. Gaylord Nelson, speech drafts re federal enforcement conference, 1969, Gaylord Nelson Papers, Subseries John Heritage, MSS 1020, box 131, folder 11, Wisconsin Historical Society Archives.

46. Nelson, draft statement for the enforcement conference. The sentence about him being described as a fanatic was crossed out and didn't appear in the final draft. Ibid.

47. John Wigren, letter to editor, in the *Minneapolis Star,* June 9, 1969. Ibid.

48. Fred Lee, letter regarding consulting for Reserve, Correspondence re Stoddard Report and others, 117.G.17.5, box II-4, Minnesota Attorney General, Natural Resources Division, Reserve Mining Company Files, State Archives, Minnesota Historical Society.

49. Reserve press release, August 12, 1970, in Public Relations Reserve, Stoddard Report Files, 117.G.17.9, box II-8, Minnesota Attorney General, Natural Resources Division, Reserve Mining Company Files, State Archives, Minnesota Historical Society.

50. Beeton testimony in Summary of the Testimony of Witnesses in the Reserve Mining Company vs. Minnesota Pollution Control Agency, June 22–August 5, 1970, 117.G.17.2F, Minnesota Attorney General, Natural Resources Division, Reserve Mining Company Files, State Archives, Minnesota Historical Society, 56–57.

51. Beeton testimony, ibid., 56.

52. Ibid., 15–16.

53. Kenneth Haley testimony, ibid., 10.

54. Ibid.

55. Fred Lee testimony, ibid., 23–25.

56. Fred Lee, Report on Water Quality, 1970, Reserve Exhibits 149–153, 117.G.17.6F, Minnesota Attorney General, Natural Resources Division, Reserve Mining Company Files, State Archives, Minnesota Historical Society, 8, 17–19.

57. Fred Lee to K. M. Haley of Reserve, April 27, 1970, in Letters and Memos, Correspondence 1954–1979, 117.G.17.9, Minnesota Attorney General, Natural Resources Division, Reserve Mining Company Files, State Archives, Minnesota Historical Society.

58. Memo, J. C. Gay to K. M. Haley, January 23, 1970, ibid.

59. Carter, "Pollution and Public Health," 33. Peter Schilling, "Hard-Fought United States vs. Reserve Mining Changed Environmentalism," February 15, 2013, *MinnPost,* https://www.minnpost.com/minnesota-history/2013/02/hard-fought-united-states-vs-reserve-mining-changed-environmentalism. Thomas R. Huffman, "Enemies of the People: Asbestos and the Reserve Mining Trial," *Minnesota History* 59, no. 7 (2005): 292–306.

60. Michael E. Berndt and William C. Brice, "The Origins of Public Concern with Taconite and Human Health: Reserve Mining and the Asbestos Case," *Regulatory Toxicology and Pharmacology* 52, no. 1 suppl. (2008): S31–39.

61. Huffman, "Enemies of the People," 296.

62. Ibid.

63. Ibid.

64. Carter, "Pollution and Public Health," 32.

65. Huffman, "Enemies of the People," 301. Carter, "Pollution and Public Health," 35.

66. EPA, Office of Enforcement and General Counsel, Studies Regarding the Effect of the Reserve Mining Company Discharge on Lake Superior, May 2, 1973, 4–5, 10, 14. Available at National Service Center for Environmental Publications, https://nepis.epa.gov.

67. Judge Devitt, comments in Bad Faith Findings Packet, 1974–1976, 127.I.5.2F IV-55, Minnesota Attorney General, Natural Resources Division, Reserve Mining Company Files, State Archives, Minnesota Historical Society.

68. U.S. District Judge Miles W. Lord, opinion, quoted in Carter, "Pollution and Public Health," 32.

69. Huffman, "Enemies of the People," 296. "Taconite Stirs Burden of Proof Debate," *Chemical Engineering News Archive* 52, no. 48 (1974): 14. In its ruling the court said that "plaintiffs have failed to prove that a demonstrable health hazard exists," ibid. Senator Gaylord Nelson responded by trying to make precaution part of the law, shifting the burden of proof in environmental litigation from the plaintiff to the defendant. His effort failed. Schilling, "Hard-Fought United States vs. Reserve Mining."

70. Michael Egan, "Communicating Knowledge: The Swedish Mercury Group and Vernacular Science, 1965–1972," in *New Natures: Joining Environmental History with Science and Technology Studies,* ed. Michael Egan, Dolly Jør-

gensen, Finn Arne Jørgensen, and Sara B. Pritchard (Pittsburgh: University of Pittsburgh Press, 2013), 110.

71. Stephen Bocking, "Dirty Lakes, Shattered Consensus," review of Terence Kehoe, *Cleaning Up the Great Lakes,* H-Environment, H-Net Reviews (November 1999), https://www.h-net.org/reviews/showpdf.php?id=3597.

Chapter Six. Mining, Toxics, and Environmental Justice for the Anishinaabe

1. While this chapter was in process, the mining company announced that it was suspending operations at the site, claiming concerns that the U.S. Environmental Protection Agency might have tried to block construction of the mine under provisions of the Clean Water Act. Susan Hedman, regional EPA administrator, denied any such intention. Notice of Suspension of Mining Operation, Timothy Myers (GTAC) to Ms. Ann Coakley, Wisconsin Department of Natural Resources, March 24, 2015.

2. The literature on Native American communities and environmental justice includes: Darren J. Ranco et al., "Environmental Justice, American Indians and the Cultural Dilemma: Developing Environmental Management for Tribal Health and Well-Being," *Environmental Justice* 4, no. 4 (2011): 221–230. Kyle Powys Whyte, "Environmental Justice in Native America," *Environmental Justice* 4, no. 4 (2011): 185–186. Arn Keeling and John Sandlos, "Environmental Justice Goes Underground? Historical Notes from Canada's Northern Mining Frontier," *Environmental Justice* 2, no. 3 (2009): 117–125. Jamie Vickery and Lori M. Hunter, "Native Americans: Where in Environmental Justice Theory and Research?" *Society and Natural Resources* 29, no. 1 (2014): 36–52. Natalia Ilyniak, "Mercury Poisoning in Grassy Narrows: Environmental Injustice, Colonialism, and Capitalist Expansion in Canada," *McGill Sociological Review* 4 (2014): 43–66. Catherine O'Neill, "Variable Justice: Environmental Standards, Contaminated Fish, and 'Acceptable' Risk to Native Peoples," *Stanford Environmental Law Journal* 19 (2000): 3. Stuart Harris and Barbara Harper, "A Method for Tribal Environmental Justice Analysis," *Environmental Justice* 4, no. 4 (2011): 231–237. James M. Grijalva, "Self-Determining Environmental Justice for Native America," *Environmental Justice* 4, no. 4 (2011): 187–192. George Lipsitz, "Walleye Warriors and White Identities: Native Americans' Treaty Rights, Composite Identities and Social Movements," *Ethnic and Racial Studies* 31, no. 1 (2008): 101–122. Dana E. Powell, "Technologies of Existence: The Indigenous Environmental Justice Movement," *Development* 49, no. 3 (2006): 125–132.

3. United Nations Ramsar Convention, "USA Names Lake Superior Bog Complex," http://www.ramsar.org/news/usa-names-lake-superior-bog -complex (accessed March 9, 2012). For additional information, see the Ramsar Sites Information Service at https://rsis.ramsar.org/ris/2001.

4. Patty Loew, *Indian Nations of Wisconsin: Histories of Endurance and Renewal,* 2nd ed. (Madison: Wisconsin Historical Society Press, 2013), 60 and 62. For analysis of the importance of wild rice for the Ojibwe people, see Thomas Vennum, *Wild Rice and the Ojibwe People* (St. Paul: Minnesota Historical Society Press, 1988).

5. Jon Sandlos and Arn Keeling, "Claiming the New North: Development and Colonialism at the Pine Point Mine, Northwest Territories, Canada," *Environment and History* 18, no. 1 (2012): 5–34.

6. Rebecca Kemble, "The Walker Regime Pushes for Controversial Mining Law," *Progressive,* October 26, 2011.

7. Tom Fitz, "The Ironwood Iron Formation of the Penokee Range," *Wisconsin People and Ideas* (Spring 2012): 33–39.

8. U.S. Steel had negotiated with the Nature Conservancy to sell it the mineral rights under condition that the land be maintained for logging but mining not be allowed (thus reducing potential competition for U.S. Steel if steel prices rose enough to make mining the deposit profitable). But in the final stages of negotiation, U.S. Steel pulled out of the deal. Later, RGGS Land and Minerals, Ltd., of Houston, Texas, and LaPointe Mining Company in Minnesota, acquired the mineral rights. Interview with Matt Dallman, director of conservation, the Nature Conservancy, Wisconsin, May 2011.

9. The CEO of GTAC is Chris Cline, who has promoted longwall coal mining in Illinois, where his operations have been cited fifty-three times over three years for violating water quality standards; see Mary Annette Pember, "Gogebic Taconite and the Penokee Hills: The Battle Rages On," *Indian Country Today,* January 28, 2011. For Clean Water Act violations, see Al Gedicks, "The Fight Against Wisconsin's Iron Mine," *Wisconsin Resources News,* April 2013.

10. Pember, "Gogebic Taconite and the Penokee Hills." Fred W. Kohlmeyer, "Pioneering with Taconite," *Minnesota History* 39, no. 4 (1964): 163–164.

11. Al Gedicks and Dave Blouin, "Science and Facts Show a Need for Tight Regulation of Taconite Mining," *Duluth News Tribune,* February 13, 2013.

12. Fitz, "Ironwood Iron Formation," 33–39.

13. John Riley discusses the Canadian construction boom and its associated rise in enormous quarries and mines in *The Once and Future Great Lakes Coun-*

try: An Ecological History (Montreal and Kingston: McGill-Queen's University Press, 2013).

14. Cory McDonald et al., "Taconite Iron Mining in Wisconsin: A Review," Wisconsin Department of Natural Resources Report, December 19, 2013, 35.

15. Danielle Kaeding, "Owners of Former GTAC Site Plan Mineral Study with USGS," Wisconsin Public Radio News, October 16, 2015.

16. Elizabeth M. Allen et al., "Mortality Experience Among Minnesota Taconite Mining Industry Workers," *Occupational and Environmental Medicine* (2014): doi:10.1136/oemed-2013-102000. Seok Jo Kim et al., "Asbestos-Induced Gastrointestinal Cancer: An Update," *Journal of Gastrointestinal and Digestive System* 3, no. 3 (2013): 135. B. S. Levy et al., "Investigating Possible Effects of Asbestos in City Water: Surveillance of Gastrointestinal Cancer Incidence in Duluth, Minnesota," *American Journal of Epidemiology* 103, no. 4 (1976): 362–368. James R. Millette et al., "Asbestos in Water Supplies of the United States," *Environmental Health Perspectives* 53 (1983): 45. Minnesota Cancer Surveillance System, Minnesota Department of Health, MCSS Epidemiology Report 97:1, September 1997.

17. Fitz, "Ironwood Iron Formation of the Penokee Range."

18. Ibid.

19. Nancy Schuldt, "What Happens in the Headwaters: Mining Impacts in the St. Louis River Watershed," page 34 in *Proceedings of the St. Louis River Estuary Summit* (February 2013), The Lake Superior National Estuarine Research Reserve, Superior, WI LSNERR Document number: 2013-R-1003; http://lsnerr.uwex.edu/Docs/2013ScienceSummit.pdf.

20. Paula Maccabee, attorney for WaterLegacy, to Commissioner Paul Aasen, Minnesota Pollution Control Agency, March 10, 2011, available at http://waterlegacy.org/sites/default/files/Regulation-Enforcement/Dunka/Dunka MineComment3-10-11.pdf. Tom Meersman, "Runoff from Old Mines Raises Fears," *Minneapolis Star Tribune*, October 1, 2010.

21. Meersman, "Runoff from Old Mines Raises Fears."

22. Nancy Schuldt, quoted in press release, Iron County Citizen's Forum, October 14, 2013; available online at http://woodsperson.blogspot.com/2013/10/minnesota-iron-range-environmental.html. The Minnesota Pollution Control Agency is currently in the middle of a controversial effort to revise the sulfate standard based on the chemistry of individual water bodies. For details, see https://www.pca.state.mn.us/water/protecting-wild-rice-waters.

23. John Myers, "Sulfate Standard Rolled Back More in House Environment Bill," *Duluth News Tribune*, March 22, 2011.

24. Barbara C. Scudder et al., "Mercury in Fish, Bed Sediment, and Water from

Streams Across the United States, 1998–2005," U.S. Geological Survey Scientific Investigations Report 2009–5109.

25. Patricia McCann, "Mercury Levels in Blood from Newborns in Lake Superior Basin," Minnesota Department of Health (MDH) Fish Consumption Advisory Program and MDH Public Health Laboratory, GLNPO ID 2007–942, November 30, 2011, available at http://www.health.state.mn.us/divs/eh/hazardous/topics/studies/glnpo.pdf.

26. M. E. Berndt, "Mercury and Mining in Minnesota: Minerals Coordinating Committee Final Report," Minnesota Department of Natural Resources, October 15, 2003, available at http://files.dnr.state.mn.us/lands_minerals/mercuryandmining.pdf. Michael Berndt and John Engesser, "Mercury Transport in Taconite Processing Facilities: (I) Release and Capture During Induration," *Iron Ore Cooperative Research Final Report,* August 15, 2005, available at http://files.dnr.state.mn.us/lands_minerals/reclamation/Berndtand Engesser2005a.pdf. Michael Berndt and Travis Bavin, "On the Cycling of Sulfur and Mercury in the St. Louis River Watershed, Northeastern Minnesota," Environmental and Natural Trust Fund, Minnesota Department of Natural Resources, August 15, 2012, available at https://gis.lic.wisc.edu/wwwlicgf/glifwc/PolyMet/FEIS/reference/Berndt%20and%20Bavin%20 2012b.pdf. Michael Berndt and Travis Bavin, "Sulfate and Mercury Chemistry of the St. Louis in Northeastern Minnesota," Minnesota Department of Natural Resources, 2009, available at https://d-commons.d.umn .edu/handle/10792/2541.

27. Patty Loew, *Indian Nations of Wisconsin: Histories of Endurance and Renewal* (Madison: Wisconsin Historical Society Press, 2013).

28. Zoltán Grossman, "Unlikely Alliances: Treaty Conflicts and Environmental Cooperation Between Native American and Rural White Communities," *American Indian Culture and Research Journal* 29, no. 4 (2005): 21–43.

29. Reynolds, "Native American Water Ethic," 154.

30. Wisconsin (State) Legislature, Senate Bill 3, 1997, Wisconsin Act 171, http://docs.legis.wisconsin.gov/1997/related/acts/171.

31. Zoltán Grossman, "Chippewa Block Acid Shipments," *Progressive,* October 1, 1996.

32. "Mine Backers Drill with Big Cash to Ease Regulations," *Wisconsin Democracy Campaign,* http://www.wisdc.org/pr012813.php, accessed August 5, 2013.

33. Senate Bill 3, 1997, Wisconsin Act 171. According to journalist Rebecca Kemble, at least $1 million was donated to mining committee members, and $15.6 million was given by "pro-mining interests to Governor Walker and

other state legislators, outspending groups opposed to the measure 610 to 1" (Rebecca Kemble, "Bad River Chippewa Take a Stand Against Walker and Mining," *Progressive*, January 28, 2013). In August 2014, when evidence emerged that the mining company Gogebic Taconite had secretly donated $700,000 to a political group helping Governor Scott Walker win election, the *New York Times* editorialized on the apparent corruption in "How to Buy a Mine in Wisconsin: Did Gov. Scott Walker Violate Campaign Laws?," *New York Times*, August 31, 2014.

34. Rebecca Kemble, "Walker's Colossal Giveaway," *Progressive*, March 5, 2013.

35. Lee Berquist, "Decision Puts Water Quality in Tribe's Hands; Sokaogon Can Set Standard near Mine," *Milwaukee Journal Sentinel*, June 4, 2002. "Walker Pushes Mining Co.'s Bill, Despite Tribe's Objections," *Progressive*, March 5, 2013.

36. Lee Bergquist, "EPA Disputes Gogebic's Fears of Agency Blocking Mine," *Milwaukee Journal Sentinel*, March 6, 2015. Because iron prices are always reported in $/metric ton, not imperial ton, I have used those measurements here.

37. Timothy LeCain, personal communication. Quotations from LeCain in the following paragraphs are also from this communication.

38. Production figures from http://en.wikipedia.org/wiki/List_of_countries _by_iron_ore_production (accessed February 3, 2017). Richard White, "Incommensurate Measures: Nature, History, and Economics," keynote lecture at conference on "Resources: Endowment or Curse, Better or Worse?," February 24, 2012, Yale University, New Haven, Conn.

Chapter Seven. The Mysteries of Toxaphene and Toxic Fish

1. Melvin J. Visser, *Cold, Clear, and Deadly: Unraveling a Toxic Legacy* (Lansing: Michigan State University Press, 2007).

2. H. J. de Geus et al., "Environmental Occurrence, Analysis, and Toxicology of Toxaphene Compounds," *Environmental Health Perspectives* 107, suppl. 1 (1999): 115.

3. Ronald Eisler, *Eisler's Encyclopedia of Environmentally Hazardous Priority Chemicals* (Amsterdam, Oxford: Elsevier, 2007), http://site.ebrary.com/ id/10187279.

4. Agency for Toxic Substances and Disease Registry (October 2014), www .atsdr.cdc.gov/ToxProfiles/tp94-c1-b.pdf. Visser, *Cold, Clear, and Deadly*.

5. H. M. Tyus, *Ecology and Conservation of Fishes* (Boca Raton: CRC Press, 2011).

6. Harold Mooney, *Invasive Alien Species: A New Synthesis* (Washington, D.C.: Island Press, 2005). For details on lake trout in Lake Superior, see Nancy Langston, "Resiliency and Collapse: Lake Trout, Sea Lamprey, and Fisheries Management in Lake Superior," pp. 239–262, in Lynne Heasley and Daniel Macfarlane, eds., *Border Flows: A Century of the Canadian-American Water Relationship* (Calgary: University of Calgary Press); P. A. Gilderhus and B. G. H. Johnson, "Effects of Sea Lamprey Control in the Great Lakes on Aquatic Plants, Invertebrates, and Amphibians," *Canadian Journal of Fisheries and Aquatic Sciences* 37 (1980): 1895–1905; Daniel W. Coble et al., "Lake Trout, Sea Lampreys, and Overfishing in the Upper Great Lakes: A Review and Reanalysis," *Transactions of the American Fisheries Society* 119, no. 6 (2011): 985–995; Philip M. Cook et al., "Effects of Aryl Hydrocarbon Receptor-Mediated Early Life Stage Toxicity on Lake Trout Populations in Lake Ontario During the 20th Century," *Environmental Science & Technology* 37, no. 17 (2003): 3864–3877; Charles C. Krueger, Michael L. Jones, and William W. Taylor, "Restoration of Lake Trout in the Great Lakes: Challenges and Strategies for Future Management," *Journal of Great Lakes Research* 21, suppl. 1 (1995): 547–558; A. H. Lawrie, "The Sea Lamprey in the Great Lakes," *Transactions of the American Fisheries Society* 99, no. 4 (1970): 766–775; Fred P. Meyer and Rosalie A. Schnick, "Sea Lamprey Control Techniques: Past, Present, and Future," *Journal of Great Lakes Research* 9, no. 3 (1983): 354–358; B. R. Smith and J. J. Tibbles, "Sea Lamprey (*Petromyzon marinus*) in Lakes Huron, Michigan, and Superior: History of Invasion and Control, 1936–78," *Canadian Journal of Fisheries and Aquatic Sciences* 37, no. 11 (1980): 1780–1801. For the St. Lawrence Seaway and the canal project, see Daniel Macfarlane, *Negotiating a River: Canada, the U.S., and the Creation of the St. Lawrence Seaway* (Toronto: University of Toronto Press, 2014).

7. Margaret Beattie Bogue, *Fishing the Great Lakes: An Environmental History, 1783–1933* (Madison: University of Wisconsin Press, 2000).

8. Smith and Tibbles, "Sea Lamprey."

9. Cook et al., "Effects of Aryl Hydrocarbon." Christopher Steiner, "Experts Reel in Clues to Lake Trout's Killer," *Chicago Tribune,* November 9, 2003.

10. Linda M. Campbell et al., "Hydroxylated PCBs and Other Chlorinated Phenolic Compounds in Lake Trout (*Salvelinus namaycush*) Blood Plasma From the Great Lakes Region," *Environmental Science & Technology* 37, no. 9 (2003): 1720–1725.

11. Derek C.G. Muir et al., "Bioaccumulation of Toxaphene Congeners in the Lake Superior Food Web," *Journal of Great Lakes Research* 30, no. 2 (2004): 316–340.

12. T. F. Bidleman and C. E. Olney, "Long Range Transport of Toxaphene Insecticide in the Atmosphere of the Western North Atlantic," *Nature* 257, no. 5526 (1975): 475–477. T. F. Bidleman and G. W. Patton, "Toxaphene and Other Organochlorines in Arctic Ocean Fauna: Evidence for Atmospheric Delivery," *Arctic* 42, no. 4 (1989): 307–313. T. F. Bidleman et al., "Selective Accumulation of Polychlorocamphenes in Aquatic Biota from the Canadian Arctic," *Environmental Toxicology and Chemistry* 12, no. 4 (1993): 701–709. Terry F. Bidleman and Andi Leone, "Soil-Air Relationships for Toxaphene in the Southern United States," *Environmental Toxicology and Chemistry* 23, no. 10 (2004): 2337–2342.

13. Susan T. Glassmeyer et al., "Toxaphene in Great Lakes Fish: A Temporal, Spatial, and Trophic Study," *Environmental Science & Technology* 31, no. 1 (1997): 84–88.

14. Susan T. Glassmeyer, Kenneth A. Brice, and Ronald A. Hites, "Atmospheric Concentrations of Toxaphene on the Coast of Lake Superior," *Journal of Great Lakes Research* 25, no. 3 (1999): 492–499. Susan T. Glassmeyer, David S. De Vault, and Ronald A. Hites, "Rates at Which Toxaphene Concentrations Decrease in Lake Trout from the Great Lakes," *Environmental Science & Technology* 34, no. 9 (2000): 1851–1855.

15. Madeline Fischer, "The Catch," *GROW Magazine,* Summer 2010, http:// grow.cals.wisc.edu/health/the-catch.

16. Catherine O'Neill, "Variable Justice: Environmental Standards, Contaminated Fish, and 'Acceptable' Risk to Native Peoples," *Stanford Environmental Law Journal* 19 (2000): 3.

17. Nancy Langston, *Toxic Bodies: Hormone Disruptors and the Legacy of DES* (New Haven: Yale University Press, 2010), 5–7.

18. U.S. Environmental Protection Agency, "Fish Consumption in the Great Lakes: Critical Contaminants in Fish," http://www.great-lakes.net/human health/fish/, accessed July 10, 2016.

19. Brian Fagan, *Elixir: A History of Water and Humankind* (New York: Bloomsbury Publishing USA, 2011), 202–210.

20. Leslie Tomory, "The Environmental History of the Early British Gas Industry, 1812–1830," *Environmental History* 17, no. 1 (2012): 29–54.

21. W. R. MacCallum and J. H. Selgeby, "Lake Superior Revisited 1984," *Canadian Journal of Fisheries and Aquatic Sciences* 44, no. S2 (1987): 23–36.

22. L. M. Fisher, "Health Service Reports 1927," *Public Health Reports* 42,

no. 37 (1927). C. A. Perry, "Studies Relative to the Significance of the Present Oyster Score," *American Journal of Epidemiology* 8, no. 5 (1928): 694–722.

23. T. Tsuda et al., "Minamata Disease: Catastrophic Poisoning Due to a Failed Public Health Response," *Journal of Public Health Policy* 30, no. 1 (2009): 54–67.

24. Brett L. Walker, *Toxic Archipelago: A History of Industrial Disease in Japan* (Seattle: University of Washington Press, 2009), 145.

25. C. W. Nicol, "Minamata: A Saga of Suffering," *Japan Times,* October 2, 2012.

26. Sören Jensen, "The PCB Story," *Ambio* 1, no. 4 (1972): 123–131.

27. Eric Francis, "Conspiracy of Silence," *Sierra Magazine,* September/October 1994.

28. U.S. Environmental Protection Agency, National Listing of Fish Advisories 2011, https://www.epa.gov/fish-tech/national-listing-fish-advisories -questions-and-answers-2011, accessed July 10, 2016.

29. Erik Stokstad, "Salmon Survey Stokes Debate About Farmed Fish," *Science* 303, no. 5655 (2004): 154. Juha Pekkanen, "Risk-Benefit Analysis of Eating Farmed Salmon," *Science* 305, no. 5683 (2004): 476–477. Jeffery A. Foran and David VanderPloeg, "Consumption Advisories for Sport Fish in the Great Lakes Basin: Jurisdictional Inconsistencies," *Journal of Great Lakes Research* 15, no. 3 (1989): 476–485. J. Burger et al., "Science, Policy, Stakeholders, and Fish Consumption Advisories: Developing a Fish Fact Sheet for the Savannah River," *Environmental Management* 27, no. 4 (2001): 501–514.

30. Becky Mansfield, "Gendered Biopolitics of Public Health: Regulation and Discipline in Seafood Consumption Advisories," *Environment and Planning D: Society and Space* 30, no. 4 (2012): 588–602.

31. Langston, *Toxic Bodies.*

32. Mansfield, "Gendered Biopolitics," 592.

33. European Food Safety Authority, "EFSA Provides Advice on the Safety and Nutritional Contribution of Wild and Farmed Fish," July 4, 2005, www.efsa .europa.eu/en/press/news/contam05070. Mansfield, "Gendered Biopolitics of Public Health," 599.

34. O'Neill, "Variable Justice," 10, 27.

35. Ibid., 45–50.

36. Federal Advisory Committee to the U.S. Environmental Protection Agency, "Fish Consumption and Environmental Justice: A Report Developed from the National Environmental Justice Advisory Council Meeting of December 3–6, 2001" (EPA: November 2002), 13.

37. Valoree Gagnon, "Fish Contaminants Through the Tribal Perspective: An Ethnography of the Keweenaw Bay Indian Community's Tribal Fish Harvest" (MS thesis, Michigan Technological University, 2011). For more on fish advisories, see Valoree Gagnon, "Environmental Justice for Seven Generations: An Institutional Ethnography of Fish, Risk, and Health in the Lake Superior Toxic Riskscape" (PhD diss., Michigan Technological University 2016).

38. Paul Rauber, "Fishing for Life," *Sierra Magazine,* November/December 2001.

39. Federal Advisory Committee, "Fish Consumption and Environmental Justice."

40. Fischer, "Catch."

Chapter Eight. The Great Lakes Water Quality Agreements

1. D. W. Schindler, "Lakes as Sentinels and Integrators for the Effects of Climate Change on Watersheds, Airsheds, and Landscapes," *Limnology and Oceanography* 54, no. 6 (2009): 2349. For modern accounts of the changing Great Lakes, see Jerry Dennis, *The Living Great Lakes: Searching for the Heart of the Inland Seas* (New York: St. Martin's Griffin, 2004), and William Ashworth, *Great Lakes Journey: A New Look at America's Freshwater Coast* (Detroit: Wayne State University Press, 2003). For Great Lakes environmental history, see James Feldman and Lynne Heasley, "Recentering North American Environmental History: Pedagogy and Scholarship in the Great Lakes Region," *Environmental History* 12, no. 4 (2007): 951–958.

2. Doug Stewart, "Will This Lake Stay Superior?" *National Wildlife Federation Magazine,* August 1, 1993.

3. Frank Quinn, "The Evolution of Federal Water Policy," *Canadian Water Resources Journal* 10, no. 4 (1985): 21–33. A key early test of pollution and the Boundary Waters Treaty came with the smelter smoke case in the late 1920s that John Wirth discusses in *Smelter Smoke in North America: The Politics of Transborder Pollution* (Lawrence: University Press of Kansas, 2000).

4. International Joint Commission, "A Guide to the GLWQA," www.ijc.org/ en/activitiesX/consultations/glwqa/guide_3.php. International Joint Commission, Boundary Waters Treaty, www.ijc.org/en_/BWT (accessed February 3, 2017).

5. Joseph T. and Alan M. Schwartz, "The Changing Environmental Role of the Canada–United States International Joint Commission," *Environmental History Review* 8, no. 3 (1984): 236–251. K. Harrington-Hughes, "Great

Lakes Water Quality: A Progress Report," *Journal of the Water Pollution Control Federation* (August 1978): 1886–1888.

6. Jockel and Schwartz, "Changing Environmental Role," 242. Harrington-Hughes, "Great Lakes Water Quality," 1886.

7. IJC, "Guide to the GLWQA."

8. Ibid.

9. Jockel and Schwartz, "Changing Environmental Role," 244.

10. International Joint Commission, *First Biennial Report Under the Great Lakes Water Quality Agreement of 1978* (1982), 6, 3. All the biennial reports are available at the International Joint Commission's website http://www.ijc .org/en_/Biennial_Reports.

11. Harrington-Hughes, "Great Lakes Water Quality," 1887.

12. International Joint Commission, *Second Biennial Report Under the Great Lakes Water Quality Agreement of 1978* (1984), 4.

13. Ibid., 4.

14. Ibid., 3.

15. International Joint Commission, *Seventh Biennial Report Under the Great Lakes Water Quality Agreement of 1978* (1994), 7.

16. International Joint Commission, *Third Biennial Report Under the Great Lakes Water Quality Agreement of 1978* (1986), 1–2.

17. Michael Egan, "History for a Sustainable Future," blog post at https:// eganhistory.com/. IJC, *Third Biennial Report*, 20.

18. Ludwig has written extensively on this; see "Contaminants Effected Widespread Changes of Great Lakes Populations and Communities," *Ecological Applications* 6 no. 3 (1996): 962–965. James Ludwig, "Let's Not Confuse Ecosystem Approaches," *FOCUS IJC* 22, no. 1 (March/April 1997). James P. Ludwig, *The Dismal State of the Great Lakes* (Bloomington Ind.: Xlibris Corporation, 2013). James Ludwig, "Has the Ecosystem Approach Improved or Diminished Sound Great Lakes Management?" 2016 draft of paper in possession of the author.

19. IJC, *Third Biennial Report*, 39. William Ashworth, *The Late, Great Lakes: An Environmental History* (Detroit: Wayne State University Press, 1987).

20. Ludwig, "Has the Ecosystem Approach Improved or Diminished?"

21. IJC, "Guide to the GLWQA."

22. Lake Superior Binational Program, *Lake Superior Lakewide Management Plan 2000* (Chicago: Lake Superior Binational Program, 2000). The EPA quote is from International Joint Commission, *Fourth Biennial Report Under the Great Lakes Water Quality Agreement of 1978* (1989), 13. Ludwig quotes from "Let's Not Confuse Ecosystem Approaches."

23. Lake Superior Binational Program, "Lake Superior Critical Pollutants," chap. 4 in *Lake Superior Lakewide Management Plan 2000*, i.

24. John H. Hartig and John R. Vallentyne, "Use of an Ecosystem Approach to Restore Degraded Areas of the Great Lakes," *Ambio* 18, no. 8 (1989): 423–428.

25. International Joint Commission, *Eighth Biennial Report on Great Lakes Water Quality* (1996), v, 3.

26. U.S. Department of State and U.S. EPA, *United States Response to Recommendations in the International Joint Commission's Eight Biennial Report on Great Lakes Water Quality* (September 1997), 1.

27. Hartig and Vallentyne, "Use of an Ecosystem Approach," 428.

28. National Wildlife Federation and Great Lakes United, "Petition to Classify Lake Superior as an Outstanding National Resource Water for Persistent Bioaccumulative Toxic Substances," October 25, 1994, 9, in author's possession.

29. International Joint Commission, *Fifth Biennial Report Under the Great Lakes Water Quality Agreement of 1978* (1991), 9, 14.

30. Dave Dempsey, *On the Brink: The Great Lakes in the 21st Century* (Lansing: Michigan State University Press, 2004), 203. I served on the Lake Superior Binational Forum from 2010 until funding was eliminated in 2015.

31. Environment Canada, *Canada's Response to Recommendations in the Sixth Biennial Report of the International Joint Commission* (1993), 10.

32. IJC, *Eighth Biennial Report*, 9.

33. Environment Canada, *Canada's Response* (1993), 7.

34. The IJC called for immediate zero discharge yet smelters remained a key source of mercury for more than a decade. *Sixth Biennial Report Under the Great Lakes Water Quality Agreement of 1978* (1993), 8.

35. International Joint Commission, *Ninth Biennial Report on Great Lakes Water Quality: Perspective and Orientation* (1998), 32. Melvin J. Visser, *Cold, Clear, and Deadly: Unraveling a Toxic Legacy* (Lansing: Michigan State University Press, May 2007). Ludwig, "Has the Ecosystem Approach Improved or Diminished?"

36. "The Parties, in cooperation with Lake Superior states and provinces, establish a specific date at which no point source release of any persistent toxic substances will be permitted into Lake Superior or its tributaries." International Joint Commission, *Fifth Biennial Report on Great Lakes Water Quality* (1991). Environment Canada, *Canada's Response to the Sixth Biennial Report*, 6, 8.

37. IJC quote from *Seventh Biennial Report*, 11, 12. EPA quote from U.S.

Department of State and U.S. EPA, *United States Response to Recommenda-
tions in the International Joint Commission's Seventh Biennial Report on Great
Lakes Water Quality* (February 1994), 16. Quote from Environment Canada,
Canada's Response to the Seventh Biennial Report (October 1994), 7–8.

38. Dempsey, *On the Brink,* 206. The commission wrote: "Protracted legal
battles to remove DDT from use foreshadowed the continued struggles to
reduce environmental contaminants. The time and resources required to
document contamination and injury to establish linkages between cause
and effect has inhibited action in a public health policy. A comprehensive
approach to all persistent toxic chemicals is not only the preferred way to
protect the integrity of the ecosystem and public health, but the only effective way." IJC, *Eighth Biennial Report,* 8. In 1996, the Binational Forum did
come up with a timetable for partial reductions of several toxics in the Lake
Superior basin. Ibid., 24.

39. IJC, *Seventh Biennial Report,* 25, 22–23.

40. IJC, *Eighth Biennial Report,* 19.

41. IJC, *Ninth Biennial Report,* 5.

42. Ibid., 13.

43. International Joint Commission, *Twelfth Biennial Report on Great Lakes Water
Quality: The Challenge to Restore and Protect the Largest Body of Fresh Water in
the World* (2004). International Joint Commission, *Thirteenth Biennial Report
on Great Lakes Water Quality: The Challenge to Restore and Protect the Largest
Body of Fresh Water in the World* (2006), 3. For its fourteenth biennial report,
the commission chose to focus on Article VI.1 (a), which calls for programs
to abate, control, and prevent pollution from *municipal* sources entering the
Great Lakes System — not from persistent pollutants. International Joint Commission, *Fourteenth Biennial Report on Great Lakes Water Quality: The Challenge to Restore and Protect the Largest Body of Fresh Water in the World* (2009),
1. For a detailed explanation of how invasives got into the Great Lakes, see
Jeff Alexander, *Pandora's Locks: The Opening of the Great Lakes – St. Lawrence
Seaway* (East Lansing: Michigan State University Press, 2009), and Dan Egan,
The Death and Life of the Great Lakes (New York: Norton, 2017).

44. International Joint Commission, *Sixteenth Biennial Report on Great Lakes
Water Quality* (2013), 9.

45. Ibid., 14–15.

46. Dempsey, *On the Brink,* 207.

47. Binational Forum to Susan Hedman, EPA administrator Region 5, July 1,
2014, in author's possession.

48. As the Binational Program report in 2006 notes, "The desired properties

of many emerging contaminants which make them effective in regard to the desired uses in society are the same properties that have led to concern when they are found in the environment." Lake Superior Binational Program, *Lake Superior Lakewide Management Plan, 1990–2005: Critical Chemical Reduction Milestones* (Chicago: Great Lakes Program Office, 2006), 46.

49. Charlotte Schubert, "Burned by Flame Retardants," *Science News* 160, no. 15 (2001): 238–239. Lake Superior Binational Program, *Critical Chemical Reduction Milestones,* 46.

50. On flame retardant use, see the article "Bromine and Fire Safety" on the industry website: www.bromine-info.org/when-did-flame-retardants-start-to -be-used/, accessed July 3, 2016. For the dramatic increase in smoking after the war, see David M. Burns et al., "Cigarette Smoking Behavior in the United States," chapter 2 of "Changes in Cigarette-Related Disease Risks and Their Implications for Prevention and Control," *Smoking and Tobacco Control Monograph No. 8,* February 1997, https://cancercontrol.cancer.gov/ brp/tcrb/monographs/8/index.html (accessed February 3, 2017). Consumption of tobacco products increased from less than one pound per capita in 1915 to more than ten pounds per capita in the 1950s. Marty Ahrens, *Home Structure Fires* (Quincy, Mass.: National Fire Protection Association, September, 2015).

51. Michael Hawthorne, "CPSC Considers Ban on Fire Retardants in Household," *Chicago Tribune,* September 28, 2015.

52. R. Liepins and F. M. Pearce, "Chemistry and Toxicity of Flame Retardants for Plastics," *Environmental Health Perspectives* 17 (1976): 55.

53. Isao Watanabe, Takashi Kashimoto, and Ryo Tatsukawa, "Polybrominated Biphenyl Ethers in Marine Fish, Shellfish and River and Marine Sediments in Japan," *Chemosphere* 16, no. 10 (1987): 2389–2396. Östen Andersson and Gun Blomkvist, "Polybrominated Aromatic Pollutants Found in Fish in Sweden," *Chemosphere* 10, no. 9 (1981): 1051–1060. R. de Winter-Sorkina et al., "Brominated Flame Retardants: Occurrence, Dietary Intake and Risk Assessment," Dutch National Institute for Public Health and the Environment RIVM Report 320100002, Netherlands (2006), available at rivm.nl/en/ Documents_and_publications/Scientific/Reports/2006/juli/Brominated_ flame_retardants_occurrence_dietary_intake_and_risk_assessment. Kerstin Nylund et al., "Analysis of Some Polyhalogenated Organic Pollutants in Sediment and Sewage Sludge," *Chemosphere* 24, no. 12 (1992): 1721–1730. Ulla Sellström et al., "Polybrominated Diphenyl Ethers (PBDE) in Biological Samples from the Swedish Environment," *Chemosphere* 26, no. 9 (1993): 1703–1718. L. Asplund et al., "Organohalogen Substances in

Muscle, Egg and Blood from M74 Syndrome Affected and Non-affected Baltic Salmon (*Salmo salar*)," *Ambio* 28 (1999): 67–76. Jacob de Boer et al., "Do Flame Retardants Threaten Ocean Life?," *Nature* 394, no. 6688 (1998): 28–29.

54. D. Meironyté, A. Bergman, and K. Norén, "Analysis of Polybrominated Diphenyl Ethers in Human Milk," *Organohalogen Compounds* 35 (1998): 387–390. Kim Hooper, "Breast Milk Monitoring Programs (BMMPs): World-wide Early Warning System for Polyhalogenated POPs and for Targeting Studies in Children's Environmental Health," *Environmental Health Perspectives* 107, no. 6 (1999): 429. Andreas Sjödin et al., "Flame Retardant Exposure: Polybrominated Diphenyl Ethers in Blood from Swedish Workers," *Environmental Health Perspectives* 107, no. 8 (1999): 643. Jennifer M. Luross et al., "Spatial Distribution of Polybrominated Diphenyl Ethers and Polybrominated Biphenyls in Lake Trout from the Laurentian Great Lakes," *Chemosphere* 46, no. 5 (2002): 665–672.

55. Quoted in Schubert, "Burned by Flame Retardants."

56. Arnold Schecter et al., "Polybrominated Diphenyl Ethers (PBDEs) in U.S. Mothers' Milk," *Environmental Health Perspectives* 111, no. 14 (2003): 1723–1729.

57. Martin Mittelstaedt, "Flame Retardant in Breast Milk Raises Concern," *Toronto Globe and Mail*, June 7, 2004.

58. Ibid.; John Jake Ryan and Dorothea F. K. Rawn, "The Brominated Flame Retardants, PBDEs and HBCD, in Canadian Human Milk Samples Collected From 1992 to 2005: Concentrations and Trends," *Environment International* 70 (2014): doi:10.1016/j.envint.2014.04.020.

59. Michigan Network for Children's Environmental Health, "PBDEs: A Toxic Threat to the Great Lakes" (2010), 7, in author's possession.

60. Ibid., 3.

61. Edna Francisco, "A Smoldering Issue," *Wisconsin Natural Resources Magazine* (2003).

62. Michigan Network for Children's Environmental Health, "PBDEs: A Toxic Threat," 8. Sergei M. Chernyak et al., "Time Trends (1983–1999) for Organochlorines and Polybrominated Diphenyl Ethers in Rainbow Smelt (*Osmerus mordax*) from Lakes Michigan, Huron, and Superior, USA," *Environmental Toxicology and Chemistry* 24, no. 7 (2005): 1632–1641. Hawthorne, "CPSC Considers Ban on Fire Retardants in Household."

63. Ling Yan Zhu and Ronald A. Hites, "Temporal Trends and Spatial Distributions of Brominated Flame Retardants in Archived Fishes from the Great Lakes," *Environmental Science & Technology* 38, no. 10 (2004): 2779–2784.

Lake Superior Binational Program, *Critical Chemical Reduction Milestones,* 40.

64. Schubert, "Burned by Flame Retardants." Susan P. Porterfield, "Vulnerability of the Developing Brain to Thyroid Abnormalities: Environmental Insults to the Thyroid System," *Environmental Health Perspectives* 102, suppl. 2 (1994): 125. "Playing with Fire," *Chicago Tribune,* May 6, 2012.

65. Patrick Callahan and Sam Roe, "Fear Fans Flames for Chemical Makers," *Chicago Tribune,* May 6, 2012.

66. Poul Harremoës et al., *Late Lessons from Early Warnings: The Precautionary Principle, 1896–2000* (Copenhagen: Citeseer, 2001), 129–130.

67. "Playing with Fire."

68. Ibid.; Harremoës et al., *Late Lessons from Early Warnings,* 131.

69. Ibid., 129–130.

70. "Playing with Fire." Alissa Cordner, *Toxic Safety* (New York: Columbia University Press, 2016).

Chapter Nine. Climate Change, Contaminants, and the Future of Lake Superior

1. John Burns, *Paradise Lost? Climate Change in the North Woods* (Madison: University of Wisconsin, Center for Biology Education, 2007), 3.

2. A. Huff and A. Thomas, "Lake Superior Climate Change Impacts and Adaptation," prepared for the Lake Superior Lakewide Action and Management Plan—Superior Work Group (2014), available at http://www.epa.gov/glnpo/lakesuperior/index.html.

3. Ibid., 9, 15, 31.

4. Ibid., 33.

5. Jay A. Austin and Steven M. Colman, "Lake Superior Summer Water Temperatures Are Increasing More Rapidly Than Regional Air Temperatures: A Positive Ice-Albedo Feedback," *Geophysical Research Letters* 34, no. 6 (2007): doi:10.1029/2006gl029021.

6. Huff and Thomas, "Lake Superior Climate Change," 34, 38.

7. Ibid., 9.

8. Ibid., 26.

9. Ibid., 59, 71.

10. Ibid., 66.

11. Stephen J. Pyne, *Awful Splendour: A Fire History of Canada* (Vancouver: University of British Columbia Press, 2011). Ben Bond-Lamberty et al., "Fire

as the Dominant Driver of Central Canadian Boreal Forest Carbon Balance," *Nature* 450, no. 7166 (2007): 89–92.

12. Robie W. Macdonald et al., "How Will Global Climate Change Affect Risks From Long-Range Transport of Persistent Organic Pollutants?," *Human and Ecological Risk Assessment* 9, no. 3 (2003): 643–660.

13. Pamela D. Noyes et al., "The Toxicology of Climate Change: Environmental Contaminants in a Warming World," *Environment International* 35, no. 6 (2009): 971–986.

14. R. W. Macdonald, T. Harner, and J. Fyfe, "Recent Climate Change in the Arctic and Its Impact on Contaminant Pathways and Interpretation of Temporal Trend Data," *Science of the Total Environment* 342, no. 1–3 (2005): 5–86; Jianmin Ma et al., "Revolatilization of Persistent Organic Pollutants in the Arctic Induced by Climate Change," *Nature Climate Change* 1, no. 5 (2011): 255–260.

15. Ibid., 652. Noyes et al., "Toxicology of Climate Change," 975.

16. Macdonald, Harner, and Fyfe, "Recent Climate Change in the Arctic," 54.

17. Ibid., 64.

18. David W. Schindler, "The Cumulative Effects of Climate Warming and Other Human Stresses on Canadian Freshwaters in the New Millennium," *Canadian Journal of Fisheries and Aquatic Sciences* 58, no. 1 (2001): 18–29. On zebra mussels and PCBs, see David Jude et al., "PCB Concentrations in Walleyes and Their Prey from the Saginaw River, Lake Huron: A Comparison Between 1990 and 2007," *Journal of Great Lakes Research* 36, no. 2 (2010): 267–276.

19. Noyes et al., "Toxicology of Climate Change," 980.

20. Ibid.

21. Douglas A. Donahue, Edward J. Dougherty, and Lee A. Meserve, "Influence of a Combination of Two Tetrachlorobiphenyl Congeners (PCB 47; PCB 77) on Thyroid Status, Choline Acetyltransferase (ChAT) Activity, and Short- and Long-Term Memory in 30-Day-Old Sprague-Dawley Rats," *Toxicology* 203, no. 1 (2004): 99–107. Bjørn Munro Jenssen, "Endocrine-Disrupting Chemicals and Climate Change: A Worst-Case Combination for Arctic Marine Mammals and Seabirds?," *Environmental Health Perspectives* 114 (2006): 76; Noyes et al., "Toxicology of Climate Change."

22. George M. Woodwell and F. T. Martin, "Persistence of DDT in Soils of Heavily Sprayed Forest Stands," *Science* 145, no. 3631 (1964): 481–483. Heather Newbold, *Life Stories: World-Renowned Scientists Reflect on Their Lives and on the Future of Life on Earth* (Berkeley: University of California Press, 2000).

23. Heidi N. Geisz et al., "Melting Glaciers: A Probable Source of DDT to the

Antarctic Marine Ecosystem," *Environmental Science & Technology* 42, no. 11 (2008): 3958–3962.

24. S. Jannicke Moe et al., "Combined and Interactive Effects of Global Climate Change and Toxicants on Populations and Communities," *Environmental Toxicology and Chemistry* 32, no. 1 (2013): 49–61. Noyes et al., "Toxicology of Climate Change," 976.

25. Huff and Thomas, "Lake Superior Climate Change," 53–54.

26. Moe et al., "Combined and Interactive Effects of Global Climate Change and Toxicants," 54.

27. Macdonald et al., "How Will Global Climate Change Affect Risks?"

28. Red Cliff Band of Lake Superior Chippewa Lake Superior Barrels Project, "Investigation Report Summary," October 26, 2015, at https://lakesuperior barrels.wordpress.com/2015/10/.

29. Huff and Thomas, "Lake Superior Climate Change," 77.

30. Ibid.

31. Steve Rayner, "Afterword," in Sarah Strauss and Benjamin S. Orlove, eds., *Weather, Climate, Culture* (New York: Berg, 2003), 283.

32. Roderick McIntosh, Joseph A. Tainter, and Susan Keech McIntosh, "Climate, History, and Human Action," in *The Way the Wind Blows: Climate, History, and Human Action* (New York: Columbia University Press, 2000), 3–4. James Rodger Fleming, "Global Climate Change and Human Agency," in *Intimate Universality: Local and Global Themes in the History of Weather and Climate* (Sagamore Beach: Science History Publications, 2006).

33. Donald Worster, "Climate and History: Lessons from the Great Plains," in *Earth, Air, Fire, Water: Humanistic Studies of the Environment* (Amherst: University of Massachusetts Press, 1999), 68–69.

Index

Page numbers in *italics* refer to illustrations.